Heinz Bloch
Optimized Equipment Lubrication

Also of Interest

Fluid Machinery.
Life Extension of Pumps, Gas Compressors and Drivers
Heinz Bloch, 2020
ISBN 978-3-11-067413-2, e-ISBN 978-3-11-067415-6

Compressor Technology Advances.
Beyond 2020
Hurlel Elliott, Heinz Bloch, 2021
ISBN 978-3-11-067873-4, e-ISBN 978-3-11-067876-5

Process Machinery.
Commissioning and Startup – An Essential Asset Management Activity
Fred K. Geitner, Ronald G. Eierman, 2021
ISBN 978-3-11-070097-8, e-ISBN 978-3-11-070107-4

Industrial Green Chemistry
Serge Kaliaguine, Jean-Luc Dubois (Eds.), 2021
ISBN 978-3-11-064684-9, e-ISBN 978-3-11-064685-6

Heinz Bloch

Optimized Equipment Lubrication

Conventional Lube, Oil Mist Technology, and Full Standby Protection

2nd Edition

DE GRUYTER

Author
Heinz Bloch
267 Sunnyvale East
Montgomery, TX 77356
United States of America
heinzpbloch@gmail.com

ISBN 978-3-11-074934-2
e-ISBN (PDF) 978-3-11-074944-1
e-ISBN (EPUB) 978-3-11-074955-7

Library of Congress Control Number: 2021947270

Bibliographic information published by the Deutsche Nationalbibliothek
The Deutsche Nationalbibliothek lists this publication in the Deutsche Nationalbibliografie; detailed bibliographic data are available on the Internet at http://dnb.dnb.de.

Cover image: Adobestock 301548692
Typesetting: Integra Software Services Pvt. Ltd.
Printing and binding: CPI books GmbH, Leck

www.degruyter.com

This book is dedicated to competent professionals, the willing and patient sharers of knowledge, communicators, and value-adders. We hold them in high esteem.

Preface

As of 2021, it has been estimated that only 6% of the world's printed knowledge can be downloaded from the Internet. Moreover, much of what can be downloaded from the Internet consists of dots, or "islands of knowledge."

Connecting the dots requires considerable experience, much of which comes from reading books. Fortunately, there are important and lasting side benefits from reading: By reading we learn to express our thoughts in the form of intelligible sentences, be they spoken or written.

Intelligibly structured sentences can make the case for technology and will be at the core of a discipline that is loosely defined as reliability engineering. Reliability engineers read no-nonsense books with technical content that share two common threads: (1) The writers know what they are talking about and (2) their writers remain fully focused on the targeted readership. Erroneous ideas are challenged by knowledgeable authors and not allowed to confuse the issue or task at hand.

Knowing our audience. This text deals with the implementation of cost and reliability-optimized equipment lubrication and preservation matters as they relate to a facility's physical assets. The targeted readership includes operating technicians, maintenance professionals, reliability engineers and, especially, managers at all job levels. Some readers will be working in oil refinery and process plant maintenance and repair shops; others could be working for EPC (engineering, procurement, construction) contractors, or as field mechanics, millwrights, project engineers, mid-level managers, and project executives in oil refining and other industries. All are equipment users and people whose actions influence equipment reliability.

Professionals and motivated individuals in general benefit from knowledge updates. This creates opportunities for authors and publishers; they respond to reader requests and the development of new technologies. We cite these among the motivators who led the author to work on the second edition of *Optimized Lubrication* soon after the release of the first (2019) edition. In particular, the material on oil mist technology deserved to be re-organized and expanded for effectively updating "manager-influencers."

Defusing anecdotes. The author is again emphasizing that this text reflects his own work-related facts and experiences. It is interesting to note how factual experiences often differ from anecdotes. This text defuses many anecdotes.

Feedback from our targeted readership indicates that there is ample room for improving equipment life. Whenever equipment MTBR (mean time between repairs) does not reach industry averages and/or profit margins fall short of reasonable projections, we (the author and his publisher) suggest that readers ponder over questions such as: Could failure to meet projections or expectations be rooted in issues that are not popular to pursue? Or: Could lack of success be rooted in a mere anecdote being passed down and implemented/applied after being separated from its original context? And here is one more: Could it be that management is distrustful of a reli-

https://doi.org/10.1515/9783110749441-202

ability technician's recommendations because, in the past, ineffective action steps were initiated by managers who were responding to opinions instead of facts? Solid professional employees must support their managers with facts, not opinions.

Management support is essential. Ideally and logically, value-adding asset reliability-related publications take past management hurdles into account. Relevant material will remind us that it would be a mistake to expect reliability improvements without the continuing support and cooperation of many interacting job functions. Unlike water flowing from a tap, reliability pursuits cannot be turned on and off at will. Putting it another way, equipment reliability is clearly affected by the implementation skills of the people in the trenches, so-to-speak, and by the perceptions of everyone between the lowly workers and their higher management. Interrupted reliability pursuits send the message that such pursuits are ranked somewhere between optional and meaningless. Continuity is important.

Every one of us fits in somewhere and somehow influences asset reliability. By way of an automobile analogy, the driver and maintenance technician and design engineer carry equal weight. If one of them slacks off, reliability becomes illusory. We need to apply the same logic in our plants. We must accept that our respective responsibilities overlap; here, everybody matters and fulfills a role.

How material is presented to the reader. The text highlights pertinent facts after explaining them accurately and non-judgmentally. Whenever possible, the lead-in paragraphs to a chapter are presented in the form of a Management Digest. One or more short summary sentences at the end of the chapter summarize what the reader has learned.

For accessibility, we aimed to index and properly cross-reference the material in this book. Facts are non-negotiable and, unlike opinions, will always add value. This is the author's core belief since his first exposure to high school physics in 1945 and a fully science-based apprenticeship with basic on-the-job training in 1950. This core belief was strengthened during and long after receiving formal engineering degrees in 1962 and 1964. Please keep this in mind when you read about *Optimized Equipment Lubrication: Conventional Lube, Oil Mist Technology, and Full Standby Protection.*

Heinz P. Bloch
Montgomery, TX, August 2021.

Acknowledgments

Don Ehlert, a pioneering technical man and valued colleague in the years from 1979 until his death in 2019, is the source of most photos relating to oil mist technology. The author found a few others in binders documenting his “prior life” at Exxon Research and Engineering and Exxon Chemicals (1965–1986). Many of the older illustrations originated with Lubrication Systems Inc., Houston, TX.

Virtually all recent and fully updated oil mist illustrations were contributed by Houston-based oil mist design and installation firm T.F. Hudgins, Houston, TX. Key contributors Weldon Mundy and Keith Macaluso went out of their way to provide high-quality graphics and pricing that reflects exact oil mist hardware procurement costs as of mid-2021.

Among several other sources of additional illustrations and narratives, special thanks go to my talented machinery engineering colleague Hurl Elliott. Hurl, a true multi-tasker, was often able to put his computer skills in good use. Both he and my son Kenneth Bloch, a process reliability expert in his own right, interrupted their engineering consulting tasks by responding to my last-minute requests. Hurl also managed to dig up snippets of information we had used in our years with Exxon Chemicals.

Appreciation is expressed to a highly proficient team of engineers at AESSEAL. They explained bearing housing protector seals in their many variants and their important applications. I was able to call on team members Chris Rehmann (with AESSEAL/USA before retiring in 2016), and David Amory, Global Marketing Manager with AESSEAL, Inc., Rotherham, United Kingdom.

Texas A&M University (TAMU) played an important role behind the scenes. The author’s involvement with TAMU’s International Turbomachinery and Pump Users Symposia commenced with participation (in the early 1970s), initially with TAMU Turbo and, starting in 1982, serving as one of the founding members of TAMU’s International Pump Users Symposia. We (the “we” is used to indicate close cooperation and frequent communication between author and publisher) acknowledge TAMU’s far-reaching contributions to the teaching of equipment reliability. TAMU routinely allowed the author to use the information included in my tutorials and presentations at TAMU in the decades since 1972. Portions of these later found their way into books, conference papers, and articles.

But I also benefited from attending dozens of skillfully led users’ discussion group sessions at perhaps two dozen TAMU’s Pump Symposia, and from the extensive networking contacts that resulted. Attendees of these discussion group sessions received and disseminated actionable information. Their knowledge was updated, and the attendees confirmed the status of lube-related issues and advancements made by their plants. In many instances, important statistics and feedback came directly from users in other parts of the world. In this regard, Marty Williams a highly experienced machinery engineer working for a world-scale oil refinery in Australia, deserves our special appreciation.

https://doi.org/10.1515/9783110749441-203

Contents

List of illustrations

https://doi.org/10.1515/9783110749441-205

Introduction

As indicated in its title, *Optimized Equipment Lubrication: Conventional Lube, Oil Mist Technology, and Standby Protection*, this book addresses three primary areas, which is why the text is presented in three overlapping parts A, B, and C.

Part A "Examining rolling element bearings and lubrication" provides guidance to users and manufacturers who have closely observed the shortcomings of lubricant application methods that were first used in the nineteenth century and were often carried over to today. To be fair, these carried-over methods can still be used today. Nineteenth-century methods will suffice for user companies that aim for low initial cost and are willing to accept an estimated 95% success rate at best.

Our text quickly informs and documents the means and measures implemented by today's reliability-focused users; these are owner-operators intent on choosing better lubricants and optimized application methods. They clearly see that only the best available technologies will allow them to become best-in-class performers. That said, Part "A" briefly highlights the best available lube application technology, for example, pure oil mist. The coverage in Part "A" will serve as a management overview; a more detailed discussion of oil mist lubrication will be found later in Part "B."

Part "A" briefly deals with best available grease lubrication; in this part, we also elaborate on how prominent pump manufacturers can get trapped in a cycle of propagating fundamentally flawed lube applications by merely tweaking oil rings. The benefit-to-cost ratio of "doing it right" is very substantial. However, the reasons for many manufacturers and users being content with tweaking are alluded to under the section "A subject matter expert's appeal to reliability managers." Being satisfied with tweaking lubrication methods that are inferior to the more highly ranked available methods is cheap initially, and is expensive in the long term. In Part "A" we will see why all major manufacturers of rolling element bearings give oil rings (at best) an average performance ranking on their listing from "best" to "least effective."

Part B "Fundamentals of oil mist technology" describes this mature and now widely used technology in considerable detail. With very few exceptions, favorable cost justification is obtained for plant-wide pure oil mist systems that include electric motor bearings and properly credit the maintenance cost avoidance for using oil mist on electric motors. Likewise, maintenance cost and repair avoidance savings accrue because standby machines are protected by blanketing with oil mist. When all such savings are considered in cost justification calculations, the verifiable results will likely differ from what was passed on in old assumptions.

Blanketing with oil mist has proved fully capable of protecting standby (non-running) machinery from environments where dust and moist air are present. Water vapor in ambient air is close to inevitability. Temperature gradients exist between night and day, and vapor condensation is possible in most instances. Part B also deals with optimally preserving the equipment before it is put into service or after it has been removed from operational service and is "mothballed" for long-term storage.

https://doi.org/10.1515/9783110749441-206

Because oil mist is the predominant storage preservation method used by the best and most profitable segments of modern industry, additional important details on this mature and cost-effective technology are given in Part B. The principles of preserving mining equipment and virtually all other machinery from moisture intrusion, degradation, and contamination are explained. Topics are interwoven and later again highlighted in stand-alone sections or subheadings. Our text contrasts best practice facilities against several known "average" facilities. One such facility was shocked at the cost and complexity as they much later attempted rejuvenation. The consulting expert was shocked only when he/she realized that, for some incomprehensible reason, the facility had opted not to preserve their machinery – even in the face of unknown delays in installation and commissioning. This brings to mind an European SME's (subject matter expert's) "Appeal to Management," which we want to share with our readers:

A subject matter expert's appeal to reliability managers. Several years ago, during a round of technical pump-related communications with a reliability engineer in Europe, he sent a particularly thought-provoking note. In 2021, this engineer still worked for a prominent oil refiner and, clearly, wanted to let us know his observations and concerns. Here is the essence of his message:

> Thank you for your email; I consider its advice a lesson in common sense. As subject matter experts (SMEs), we are engineers who should make decisions that reflect applicable knowledge instead of mere opinions. Our advice to management must be rooted in science, experience, well-researched and authoritative technical texts, fact-based case histories, and relevant published articles.
>
> Personally, I like to read, read again, and then ponder over what I have read. Almost everything is documented somewhere, and we can be certain there is always someone in the world who has encountered the same problems or failures. But I consider it even more important and quite worthy of our appreciation that some still unselfishly share their experience with others. I endeavor to do the same with my colleagues in other refineries and at Technical Universities in my home country.
>
> Yet, the problem I often encounter in our operating plants is that reliability engineers are expected to act as both superman and politician. There are managers whose hope it is that the reliability engineer knows everything and, thus, does not require background data, or that this professional has access to outside experts who will contribute much time and effort with neither pay nor even a "thank you" for their valuable input. Back at the plant, and with limited time and many budget constraints, what is often left are several layers of uninformed managers who tend to make decisions based on opinions rather than facts. My younger colleagues and I try to convince management that there are times when our technical expertise and factual knowledge deserve to be heard.
>
> A case in point are matters of process-pump lubrication and my colleagues have made this observation as well. It is not that the pump manufacturer does not know much about issues traceable to inexperienced specification writers, but rather that our goals differ from those of the manufacturers. Nevertheless, our work as value-adding professionals can become unnecessarily complicated whenever managers disregard what we have absorbed over the years. There are many times when we could offer wisdom beyond that displayed by the manufacturer's inexperienced technicians. As diligent readers, our reliability professionals are frequently able to explain technology-backed reasons for defect development. If allowed, we could often suggest experience-based steps that would solve problems.

Well, this book will appeal to the corporate SME who wrote the above. His comments were certainly not unique, neither were his observations limited to Europe. His candor was commendable.

On the topic of standstill protection, the technical descriptions in this book repeatedly make the point that machines left unprotected for years or even just a few months cannot be returned to safe and dependable operating condition by belatedly applying purely preservative measures. Preservation obviates restoration. Without preservation, costly restoration cannot be avoided. In all such instances, dismantling inspections will be needed, and follow-up actions will likely involve time-consuming cleaning, reconditioning, and replacement of parts or entire machines. Protecting the shareholders' assets is every employee's obligation; being informed and keeping superiors informed are first among the keys to protecting physical assets. Learning to speak truth to power will help immensely.

Part C "Full equipment standstill/standby protection" gives details on the probable degree of infant mortality, here defined as rotating equipment failing to perform within 30 days of restarting. A comparison graph will be shown for machines that were in good shape when shut down but then left unprotected, versus the same machines which upon being shut down were properly protected for long-term preservation.

Because of its demonstrated superiority in the decades since 1972, much generic information is submitted on optimum storage protection with oil mist. Details are also given on conventional preservation techniques which, however, are always more maintenance intensive than oil mist preservation. Nevertheless, the text highlights short-term protection practices as they pertain to different machine types. These, too, represent guidelines in generic form.

Also included are remarks based on consultative advice the author gave to the managers of a large oil refining complex. The refinery had experienced a sudden project cancellation; its overseas headquarters had insisted that funding be stopped immediately. After 3 years of unprotected outdoor storage, funding was restored and the situation assessed. Attempts were then made to define which equipment to dismantle and rebuild before commissioning the plant. The refinery paid dearly for not having the foresight to preserve and protect its hundreds of assets in 3 years of outdoor exposure to the elements.

Taking into account the urgent needs of lubrication-related technical work, many of the book's sections and chapters can be separately reproduced and placed in the craftsperson's toolbox. Much of the material is presented in a user-friendly format that can help reliability-technical personnel tasked with the development of checklists for field use. And please recall that "overlapping coverage" may mean that the author decided to occasionally approach the same topic from two or more different angles, so-to-speak.

Part A: **Bearings and optimally applying lubricant**

Examining rolling element bearings and traditional lubrication

Chapter 1
Making the case for upgrading

1.1 Management digest

It would be overly optimistic or even naïve to assume that all lubrication-related design decisions made by the equipment manufacturer serve the user's best long-term interests. Understandably, cost competitiveness will have been foremost on the manufacturer's mind. After all, many machines are still purchased with initial cost as the primary – if not only – criterion. These machines are your candidates for upgrading. Upgrading by paying a small incremental added charge makes far more sense than upgrading after many repeat failures. Upgrading at the specification stage is possible if the specification writer is knowledgeable. Knowledge is a quality that builds up; lack of knowledge leads to decay in every sense of the word.

1.2 What causes lubricants to degrade

It can be shown that upgrading to superior lubricants, protecting lubricants from premature degradation and improving the method by which lubricants are delivered to bearings is often both feasible and desirable. Occasionally, a "naysayer" argues that lubricants never wear out, but the naysayers are wrong. Lubricants can suffer from gradual depletion of additives, or contamination, or the effects of excessive temperatures. Water causes partitioning, essentially a separation of molecules from certain beneficial additives. Water plus dust particles cause sludge to form. Common sense tells us that issues with lubricants can render the fluid unserviceable to the point of initiating catastrophic machine failures.

Experience also indicates that manufacturers are satisfied if, in their view, "traditional" maintenance frequencies or intensities are carried out. Similarly, a vendor-manufacturer may be satisfied if, of the 100 machines delivered to their Customer X, only 95 are reaching the industry average life of, say, three operating years. Thus, in this hypothetical case, out of every 100 machines, 5 would experience avoidable failures within this time.

But suppose that, in this arbitrary example, Customer Y has all 100 of his machines exceed the industry average and they operate for 3 years before one of them needs a repair. In that case, Customer X will spend money on repairs while Customer Y has no such expenses or outlays. Chances are that Customer Y is more successful because they implemented suitable upgrades and Customer X should give thought to upgrading.

Consider this our way of claiming that our text deals with eliminating the 5% of "elusive" repeat failures. Arguing that traditional methods and practices still suffice

https://doi.org/10.1515/9783110749441-001

is a bit like pointing out that people can still get from one place to another in a 1915 Model T Ford automobile. While agreeing with that statement, we would have no difficulty explaining and accepting that a 2021 mid-size Ford automobile will better serve our low-maintenance cost and high-reliability goals.

Elusive pump failures are, in all probability, consuming a disproportionate amount of the maintenance budget. Years ago, the author compiled statistics that placed from 7% to 10% of a facility's process pumps in the frequent failure (or "bad actor") category. Usually, about 60% of the maintenance budget for the equipment category process pumps was consumed by this 7–10% low-performing population.

1.3 Cost-justifying upgrades

An empirical assessment makes the conservative assumption that a simple available upgrade measure will extend safe operating life by factors ranging from 1.1 to 1.4, that implementing two available upgrade measures would extend safe operating life by factors from perhaps 1.5 to 2.5, and that three low-cost improvement measures would move pump operating lives to multipliers in the range from 2.6 to roughly 3.3. These approximations are often used in initial cost justification calculations; they have usually yielded reasonably close results. Proceeding with upgrade plans is considered justified if payback is obtained within 18 or fewer months.

Another rule of thumb uses an exponential approach. That rule states that if a fully upgraded machine has a reliability of 1.0, then one missed upgrade will lower the reliability to 90% of 1.0 = 0.9; two missed upgrades to 90% of 0.9 = 0.81; three missed upgrades to 90% of 0.81 = 0.73; four missed upgrades to 90% of 0.73, equaling only 0.66, and so forth. We consider this elementary rule of thumb rather optimistic. Actual achieved reliability with four deficiencies is probably less than 50% of what would be achievable with better bearings, better mechanical seals, better couplings, better constant level lubricators, or whatever other upgrades are available and within reach.

Then there is a third rule of thumb worth sharing. Again, a reasonable assumption is made; a probable 20% improvement in failure avoidance, or repair cost reductions, or life extension is thought to result from each upgrade. In that case, an upgrade will move the equipment reliability from 1.0 to 1.2, a second (different) upgrade would capture $1.2^2 = 1.44$; further upgrades $1.2^3 = 1.73$, and $1.2^4 = 2.07$. The implementation of four proven upgrade measures would cause the MTBR (mean time between repairs) to be extended slightly beyond twofold. Yearly repair expenditures would be one half of what they were before; workers previously laboring on repairs would now spend time on repair avoidance tasks. Safety would go up, community goodwill would be given a boost, and so would worker morale.

Making good use of shortcut calculation is encouraged by a good management team. Good managers routinely ask responsible staffers to accept responsibility for cost justifying and advocating reliability improvements that yield rapid payback.

These employees would be encouraged to become familiar with the above three reasonably accurate shortcut calculations. In turn, these employees would accept the task of engaging in 12 management-sponsored actions and pursuits:

- *Define equipment operating capability (reliability) limits to adequately prevent lubrication-related failures*
- *Develop lubrication strategies sufficient to maintain equipment operation and availability within specified limits*
- *Prioritize detection of limit deviation and definition of response criteria according to known or anticipated failure intervals and consequences*
- *Enforce and own the policy and procedures for lubrication-related limit changes*
- *Document the approval of new, and changes to existing, lubrication-related reliability limits*
- *Establish lube application-related limit documentation and ascertain access capabilities to retain lubricant performance limit, its purpose, and its history*
- *Set expectations for upgrading equipment assigned to limit monitoring points and assist in creating effective contingency plans for maintenance deviations*
- *Track and monitor limit compliance by contractors and the company workforce members*
- *Investigate chronic limit deviations to detect and address potential constraints that might degrade business value*
- *Communicate program performance measures on a routine basis (percent in control, chronic limit deviations)*
- *Reconcile lubricant limit performance against turnaround maintenance inspection results*
- *Audit reliability limit database integrity*

In summary, we have learned that best-in-class companies have institutionalized the study and dissemination of best practice details given in this book. But we have also learned that the decision to upgrade one's method of lube application quite often depends on manufacturers' input. Likewise, the decisions are at least influenced by the ranking which experienced users assign to these applications. The following chapter explains these rankings.

Chapter 2
Fundamentals of rolling element bearings and lubricant application

2.1 Management digest

Some bearings are not ideal for some services. The same is true for lube application methods, and some are decidedly better than others. The various methods have, over the years, been listed in the order of preference by users and bearing manufacturers. One widely used ranking order is found in the Eschmann, Hasbargen, and Weigand book *Ball and Roller Bearings* [1] for oil- and grease-lubricated bearings. These three authors were for decades employed by German bearing manufacturer Kugelfischer-FAG; the three have made major contributions to relevant literature.

A few pages into this chapter, the ranking found in [2] will be reassessed. It was developed with input representing the collective experience of a worldwide Corporate Rotating Machinery Network. Its members were equipment reliability professionals who for at least two decades formally and periodically shared relevant information with the corporation's affiliates. The collected data came from fluid machines and drivers installed in oil refineries and petrochemical plants in eight industrialized countries. The collected data reinforced the author's observations, which the reader will find captured in this chapter.

2.1.1 Pumps used as examples

Next to electric motors centrifugal pumps are the machines that best assist us in understanding how and why rolling element bearings are used. This is why a typical pump cross-sectional view is shown in Fig. 2.1. On the left side of the shaft is a double-row thrust bearing; a cylindrical roller bearing is shown on the right. There exist many hundreds of bearing styles and geometries, and this text shows perhaps fewer than one-tenth of 1% of what is available in the marketplace today. That said, our chapter merely scratches the surface of the untold possibilities and combinations.

Angular contact bearings (Fig. 2.2) are configured to allow simultaneous loading in the axial and radial directions. The cage or ball separator shown in Fig. 2.2 is made of brass or bronze. Because metal cages tolerate more heat than plastic cages, and plastic cages can be damaged if in the mounting process excessive heat is applied, cage damage can result. Needless to say, shops with the correct mounting tools and a careful workforce will not overheat a bearing. Moreover, the overall performance of today's high-performance plastics (HPPs) can favor HPP cages over the old-style brass and bronze versions.

https://doi.org/10.1515/9783110749441-002

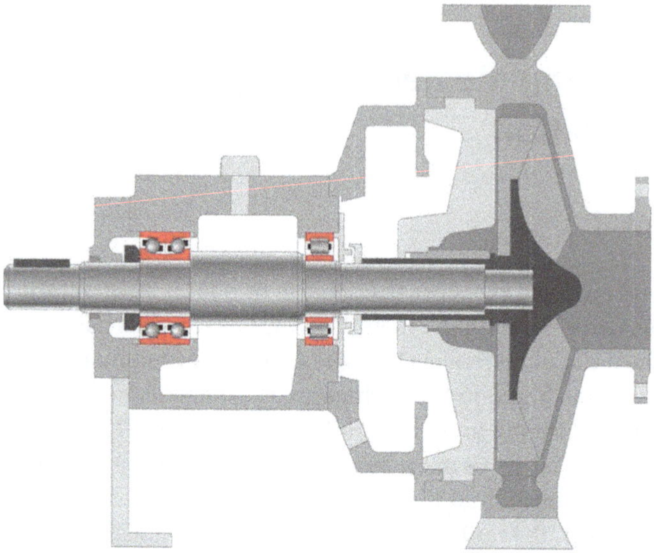

Fig. 2.1: Generic pump cross-section (source: NSK).

Working with experienced application engineers from major bearing manufacturers can prove enlightening. An astute engineer may point out to users that the generalizations found in certain industry standards may not always convey industry-leading technology. The American Petroleum Institute's prominent API-610 pump standard clearly mentions that its clauses and recommendations are minimum requirements and that users are free to use equipment that exceeds these requirements. Best-in-class owner-purchasers seek out reliability-improving components and machines.

Fig. 2.2: Angular contact bearing with "yellow metal" cage. The wide shoulder of the outer ring in an angular contact bearing is called "the back." The narrow shoulder is called "the front." (source: NSK.)

While the bearing style in Fig. 2.2 and its brass cage have long been recommended by API-610, copper-containing yellow metals (mainly brass and bronze) are not immune to "smearing" under skidding conditions. Skidding will take place in double- and triple-row bearings if the axial load acts on only one of the two or three bearings. Skidding generates heat and degrades the oil.

A bearing with a disintegrated yellow metal cage is shown in Fig. 2.3. Bearings and lubricants interact, and both should be selected with life extension in mind. Every little detail matters.

Fig. 2.3: Failed brass bearing cage in an API-compliant pump (source: HPB).

1. For best long-term trouble-free operation, reliability-focused user may opt to specify and install a bearing with equal load angles in the high thrust location and a bearing with unequal angles on the lightly loaded location.
2. Since the cages in certain types of angular contact bearings are slightly slanted, they will create a fan effect. A fan takes air from its smaller diameter intake side to its larger diameter out. If the lube application attempts to promote flow from the "fan" outlet toward its air intake, there will be the risk of inadequate oil quantities reaching the rolling elements of the bearing.
3. Visualize from Fig. 2.4 what will happen if pipe-induced stresses were to cause the pump casing to deform. Such deformation can cause bearing inner and outer rings to be out-of-parallel relative to each other. The rolling balls could then contact the rim or edge of a bearing, and the resulting localized overload stresses would cause the bearing to fail prematurely.

Angular contact bearings are shown back-to-back mounted in Fig. 2.5 (left side) and face-to-face mounted (Fig. 2.5, right side). Proper orientation is important, as is an understanding of load direction and preload effects. Proper preload avoids skidding;

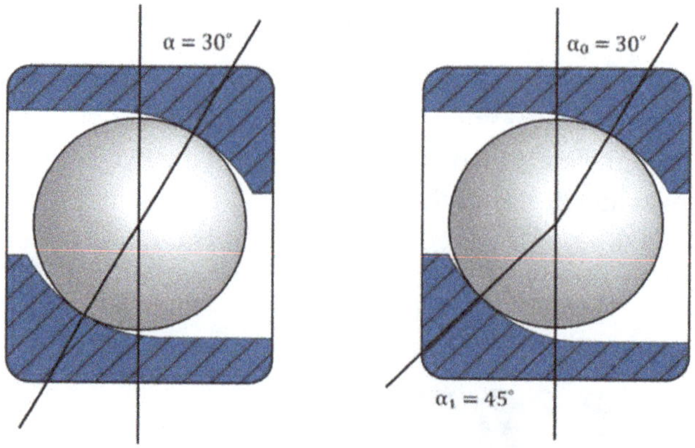

Fig. 2.4: Sets of bearings with unequal load angles (shown here in a tandem arrangement) can be problem solvers. If used as thrust bearings, they should be mounted back-to-back. The 15° bearing should be lightly loaded to reduce skidding risk (image source: Hydro, Inc., Chicago, IL).

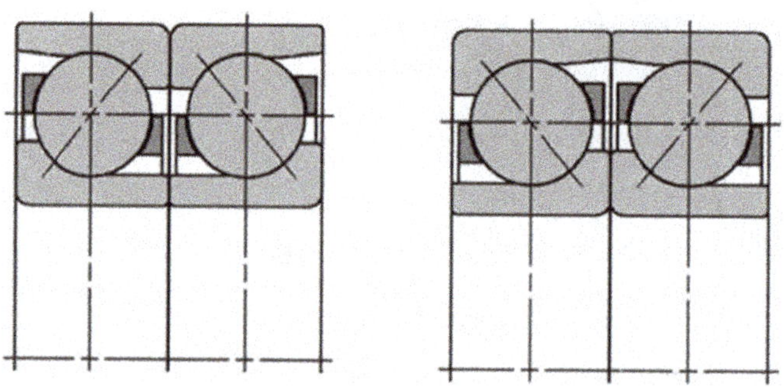

Fig. 2.5: Angular contact sets oriented back-to-back (left) and face-to-face (right) (source: NSK).

excessive preload can shorten bearing life. Different machines deserve different degrees of attention, and efforts to standardize using a "one-type-suits-all" approach are rarely (if ever) advisable [3].

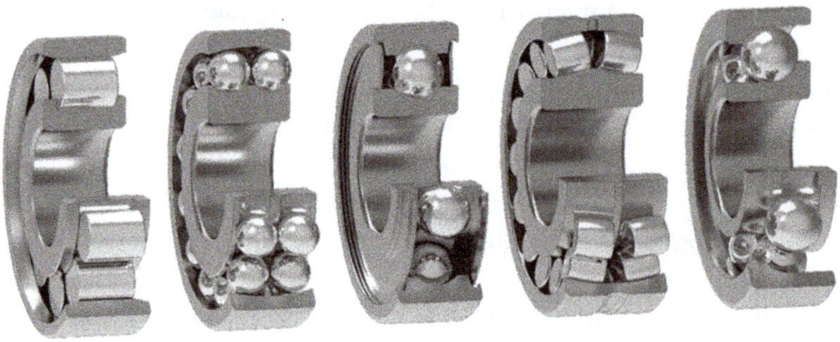

Fig. 2.6: Five of hundreds of bearing styles found in process machines (source: SKF).

Five of literally hundreds of bearing styles found in process machines are shown in Fig. 2.6. Depicted are, from left to right: A cylindrical roller bearing (typically used in high radial load situations similar to Fig. 2.1); a double-row ball bearing (with its contoured ball track in the outer ring allowing minor shaft angularity); a conventional "symmetrical" bearing with two shields to retain grease; a double-row spherical roller bearing for high loads and not quite parallel shafts; and, finally, an angular contact bearing intended for operation with high axial thrust and a lesser load acting in the radial direction. It should be noted that a load *will* be required in the radial direction; without such a load the bearing will have a shortened life.

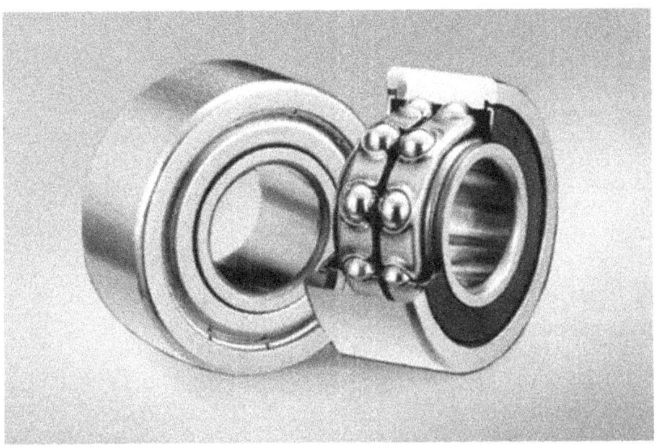

Fig. 2.7: Shielded double-row (left) and sealed double-row bearing (right).

The shielded and sealed double-row ball bearings in Fig. 2.7 find extensive use in high radial load applications [4]. Note that the conventional ball tracks in the outer ring raceway will not accommodate other than parallel inner ring bores. In other words, bearing bores must remain parallel to the bearing outer ring surfaces. Shafts cannot be at angles to bearing bore and/or outside surfaces.

Fig. 2.8: Double-row bearing with two separate inner rings.

There can be advantages to double-row designs with two separate inner rings (Fig. 2.8). Placing a thin shim between the separate inner rings accomplishes preload adjustments, if needed [5].

Fig. 2.9: A bearing with riveted cage. While perfectly acceptable for most industrial applications, riveted cages are not permitted in API-style process pumps (source: SKF).

The bearing in Fig. 2.9 incorporates a riveted cage. Riveted cages are suitable in even-temperature applications where many millions of revolutions are unlikely (skateboards, baby strollers, etc.). While perfectly suited for most industrial applications, riveted cages are not acceptable in API-style process pumps. Temperature-induced stresses can make rivet heads the weakest link in such bearings. Should one of the rivet heads "pop off," the others will soon follow. All it takes is one rivet head in the ball track and calamity can befall the user.

While at first glance the two roller bearings in Fig. 2.10 seem identical, their cage configurations differ. This is but one of hundreds of examples where input from application engineers will be of great value and payback for sound advice can be huge [6].

2.2 Failure distribution

2.2.1 Pump failure distribution

In 1986, a facility with slightly over 3,200 process pumps reported their failure distribution over a 5-year period. Then, as today, the reporting plant recognized that all failures, regardless of what machines are involved, can be attributed to one or perhaps two meaningful classifications of failure causes:

- Design defects
- Material defects

- Processing and manufacturing deficiencies
- Assembly or installation defects
- Off-design or unintended service conditions
- Maintenance deficiencies (neglect, procedures)
- Improper operation

Fig. 2.10: Two different high-capacity roller bearings (source: NSK).

The plant reporting its failure distribution also accepted as a fact that mechanical parts can fail due to one of only four cause categories: Force, reactive environment, time, and temperature. For the past five decades, we have remembered these by the acronym "FRETT" [7]. Forward-looking plants have taught "FRETT" to their field and shop maintenance workforces. "FRETT" is a theme worth explaining several times in this book.

2.2.2 Causes of bearing failures

At an oil refining facility with 3,200 process pumps and often visited by the author, the failure distribution was roughly as follows [8]:

– Design defects	6%
– Material defects	4%
– Processing and manufacturing deficiencies	8%
– Assembly or installation defects	20%
– Off-design or unintended service conditions	18%
– Maintenance deficiencies (neglect, procedures)	32%
– Improper operation	12%

Upon closer examination these statistics convey two important observations:
- Most pumps fail because of maintenance- and installation-related defects.
- Since pumps generally represent a mature product, fundamental design defects are relatively infrequent.

Reliability professionals at that facility could readily vouch for two more facts:
- Pump failure reductions are largely achieved by appropriate action of plant reliability staff and competent plant or contract maintenance work forces.
- Contractors are long-term assignees to the plant; at best-in-class facilities they receive the same training as regular employees.

2.2.3 Employee motivation matters

From the above failure distribution, it appears that the pump manufacturers rarely are to be blamed. Nevertheless, it is unfortunate that not all pump manufacturers are knowledgeable in pump failure avoidance. That is why the owners' engineers must take the lead in identifying and upgrading the "weak links." Judicious upgrading is very often possible and is generally very cost-effective. The merits of upgrading can be visualized by pondering the size of large plants.

2.2.4 Plant size is not a factor

In the mid-1980s, a chemical plant in Tennessee had over 30,000 pumps installed and a large facility near Frankfurt, Germany, reported over 20,000 pumps. However, the largest industrial pump user appeared to be a city-sized plant situated on the banks of the Rhine River south of Frankfurt. There were approximately 55,000 pumps installed at that one location alone.

US oil refineries typically operate from 600 pumps in small, to 3,600 pumps in large facilities. Among the old refineries are some that have average pump operating times of over 9 years. However, there still are some that achieve an average of only about 20 months. Some of the very good oil refineries are new, but some in the same "very good" category are relatively old. Certain bad performers belong to multi-plant owner "X" and some good performers also belong to the same owner "X." It can therefore be said that equipment age does not preclude obtaining satisfactory equipment reliability. Organizational mindsets and employee motivation are important and will make the difference.

2.2.5 Reliability-focus versus repair-focus

Experience shows that facilities with low pump mean-time-between failure (MTBF) or mean-time-between repair (MTBR) are almost always repair-focused, whereas plants with high pump MTBF are essentially reliability-focused. Repair-focused mechanics or maintenance workers see a defective part and simply replace it in kind. Repair-focused plants turn their mechanics into parts changers instead of failure analysts. Reliability-focused plants ask why the part failed, determine whether upgrading is feasible, and then determine the cost justification or economic payback achieved by implementing suitable upgrade measures. Needless to say, and always well worth repeating: Reliability-focused plants teach "FRETT" and motivate their maintenance-technical staff [8]. These staffers seek to implement every cost-justified improvement as soon as possible. Recall that Chapter 1 outlines how easy it is to investigate probable payback.

Thus, again, *why* the power end of a pump fails, and *how to avoid* failures, will be discussed in this chapter. Why the same pump model does well at one plant and does not do well at another plant will, in some cases, be described and analyzed. Pump life extension should be the overriding concern and will be the overall theme of this segment of the text.

2.3 Even elusive failures have causes

But even elusive failures have causes, and causes can be discovered. Suppose the pump bearing housing shown in Fig. 2.11 requires bearing replacement. But why did the bearings fail? Did you notice that lube oil trapped between the bearings and their respective end caps cannot escape? Trapped oil will overheat, and its carbon residue degrades both lubricant and bearing [3]). Both will degrade if drain holes are missing from the picture; insist that drain holes are provided. Unless this discrepancy is recognized and rectified by a reliability-focused owner or a competent pump repair shop (the "CPRS"), the degradation events will repeat themselves. Short bearing life will result and probably cost the owner company and its stakeholders dearly.

While replacing bearings, you might also pay attention to Fig. 2.12. Here, an oil ring shows substantial abrasion damage, and several contributing reasons likely exist. Cheap oil rings are not usually stress-relief annealed. Unless stress-relieved before finish-machining, they will become out-of-round and will slip and skip. Also, unless the driving and driven shafts are installed parallel to the true horizon, oil rings will run downhill and often contact bearing housing-internal or cast-in components. As abrasive wear takes place, the two oil rings in Fig. 2.12 will slow down, and contaminated oil will reach the bearings [9].

Both examples are among dozens which demonstrate the value of employing and retaining a well-trained work force. Some plants practice structured and well-guided maintenance efforts; moreover, good supervisors do double duty as experienced teachers. Their work execution and follow-up inspections are well planned and executed. Still, other plants are quite remiss in allocating time, brain power, and monetary resources to these essential pursuits. Also, one plant may be situated in a geographic area with an abundance of competent repair shops while another plant is not.

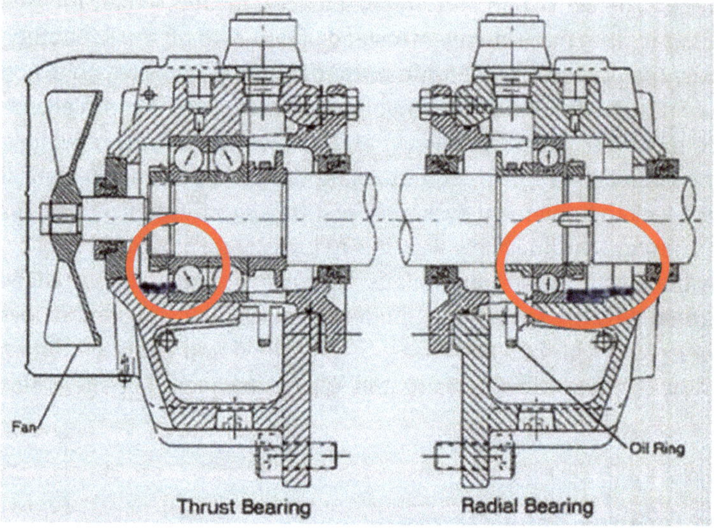

Fig. 2.11: Lube oil trapped between bearings and their respective end caps cannot escape. Consider this an elusive cause of bearing failure.

Fig. 2.12: Oil rings in new (left) and badly worn (right) condition. Cheap oil rings are not stress-relief annealed; they tend to deform and malfunction. Oil rings abrade while skipping around and contacting housing-internal surfaces [22].

2.3.1 True keys to asset performance

When the implications of Figs. 2.11 and 2.12 were pointed out to groups of maintenance and reliability technicians in four different countries, it was clear that fault-inducing details of this type were constantly being overlooked. It was also clear that organizational initiatives such as Asset Management and Operational Excellence had gained prominence, whereas attention to detail and the systematic acquisition of highly relevant in-depth knowledge remained vastly under-appreciated [9]. When it's all said and done, understanding these details is the absolute bottom line, one of the true keys to asset performance.

The other true keys to asset performance include design knowledge, maintenance knowledge, and operating knowledge. Each of these branches of knowledge needs to be implemented responsibly. There must be accountability and even-handedness in all relationships among employees and their supervisors. An ethical reward system needs to be in place for knowledge to be practiced and implemented with unerring continuity and wisdom.

But so as not to lose sight of the key words *Bearings and Lubrication* in our chapter heading, there are strong indications that fewer spare parts will be consumed by facilities which implement and inculcate the necessary knowledge in well-structured ways. Plants and facilities staffed by highly competent maintenance and reliability professionals hate having to cope with repeat failures. These professionals view every single maintenance intervention as a challenge to their desire to upgrade, or the desire to design-out the need for excessively frequent preventive maintenance actions. Highly competent maintenance and reliability professionals loathe unplanned downtime events and the need for too much maintenance. They estimate the value of upgrading by making good use of the information in Chapter 1 and see to it that weak links in the component chain are beefed up or, better yet, designed-out.

2.4 Only two root causes of failure exist

2.4.1 Repeat failures

Solid professionals agree without equivocation that all repeat failures must be ascribed either to the fact that the root cause of a failure has not been determined, or that the root cause is indeed known but remedial action is not being pursued. Highly experienced practitioners of maintenance and reliability improvement skills will never be at ease with either of these two possibilities, that is, not knowing what led to a failure, or knowing why an asset failed and doing nothing about it [10]. Competent practitioners view these challenges as personal affronts – a bit like a medical doctor losing a patient to the common cold.

2.4.2 Bearing checklist

It is intuitively evident that paying attention to the bearing issues listed next as items 1 through 13 has value. Many users have distributed them in checklist format and have incorporated them in their specifications and repair procedures:

1. Do not use filling-notch bearings (Fig. 2.13) in centrifugal pumps [11]. Replace them with dimensionally identical Conrad (center image, Fig. 2.6) or, in non-API pumps, with double-row, double inner ring, angular contact bearings (SKF Series 5300UPG, Fig. 2.8).

Fig. 2.13: Bearing with filling notches (filling slots) and riveted cages should not be used in process pumps.

2. Beware of two-piece riveted steel cage bearings (Fig. 2.13). Under certain conditions, the rivet heads prove to be the "weak link" and may pop off. Use only stamped steel, or contour-machined brass, or high-performance plastic cages in your critical pumps. Work with application engineers employed by prominent bearing manufacturers.
3. All bearings will deform under load. On dual thrust bearing sets that allow load action in both directions (i.e., axially toward the impeller and axially away from the impeller), deformation of the loaded side could cause excessive looseness – hence, skidding – of the unloaded side. This skidding may result in serious heat generation and reduced oil film thickness. Metal-to-metal contact will take place and destroy the bearing. Always select matched bearing configurations that limit or preclude skidding [5].
4. The API-610 recommended combination of two back-to-back mounted 40° angular contact bearings is generally sufficient, but will not always be the best choice for a particular pump application. Matched sets of 40° and 15° angular contact bearings, sometimes called "SKF PumPac" bearings, are designed to avoid, or greatly reduce skidding. While not a cure-all, PumPacs and/or sets of slightly pre-loaded bearings are often highly appropriate for a given service or application [6].

5. Keep in mind the possibility of using a 9000-series thrust bearing together with a 7200- or 7300-series angular contact bearing in the thrust location of certain pumps [7]. Seek factory engineering advice before finalizing your selection. Purchase your pump bearings only from companies that support their customers with good application engineers.
6. Thrust bearing axial float, that is, the total amount of movement possible between thrust bearing outer ring and bearing housing end cap, should not exceed 0.002″ (0.05 mm). Some hand-fitting or shimming may be necessary to thus limit the potentially high bearing-internal axial acceleration forces. Some pump manufacturers have allowed looseness up to 0.008″ (0.2 mm) to make their manufacturing processes less expensive [8].
7. For some applications, consider replacing old-style double-row angular contact bearings (bearings with one inner and one outer ring) with newer, Series 5300UPG, double-row angular contact bearings. These bearings have two brass cages, one outer ring and two inner rings per bearing. Note that axial clamping of the two inner rings will be necessary but that these newer double-row bearings resist skidding [4].
8. Observe allowable assembly tolerances for rolling element bearings in pumps with steel shafts:
   - Conrad, single angular contact and double-row bearings, bore-to-shaft:
   - 0.0002–0.0007" (0.005–0.018 mm) interference fit for alloy steel, but 50% less for stainless steel.
   - Bearing outside diameter-to-housing fit:
     0.0007"–0.0015" (0.02–0.04 mm) loose fit [10].

   Again, note that stainless steel shafts will require reduced bearing interference fits because stainless steels have larger coefficients of expansion.
9. Do not allow high-temperature bearing heaters for installing bearings with high-performance plastic cages. Mounting temperatures in excess of about 230 °F will weaken some plastic bearing cages.
   - Plastic cages tend to be damaged unless highly controlled mounting temperatures are maintained. Mounting by applying force to the bearing inner ring (using an arbor press) [6] is an available best practice.
   - Plastic cage degradation will not show up in conventional low-cost vibration data acquisition and analysis [9].
10. Be certain to use only precision-ground or light-preload, matched sets of thrust bearings in either back-to-back thrust or tandem thrust applications. Matched bearings must be furnished by the same bearing manufacturer. Verify precision grinding by observing that appropriate alphanumerics have been etched into the wide shoulder (usually called the "back") of the outer ring. Understand what the different code letters and alphanumerics mean. Work with a bearing manufacturer/supplier/application engineer that allows full access to application engineering departments.

11. Use radial bearings with C3 clearances in electric motors so as to accommodate thermal growth of the hotter-running bearing inner ring. Note that shear-resistant polyurea "EM" greases are preferred for electric motors in all but the highest temperature applications. For extreme high temperatures, consider aluminum-based or perfluoro-polyether (PFPE) grease formulations. Note, however, that the latter two will tolerate not even the slightest trace amounts of other greases.
12. On vertical deep well pumps, investigate column bearing (sleeve bearing) material upgrade options. Work with either the OEM or a top-rated independent repair facility.
13. Ascertain that the facility is aware of the various vulnerabilities we found in Fig. 2.14: (a) An appropriately sized oil return slot (or channel) is shown below the radical bearing. (b) Because the designer overlooked the need for an oil slot below the thrust bearing, oil could get trapped to the right of this bearing. Trapped oil tends to overheat and oxidize behind the thrust bearing unless an oil return slot is provided. (c) An oil ring that jumps out of its groove can get wedged between the tip of the limiter screw and the shaft. (e) Unless a specific constant level lubricator is shown, the vendor will likely provide the least expensive one (f) With no bearing protector seal shown, expect the cheapest "isolator" will be provided [10].

2.5 What to upgrade in process pump bearing housings

2.5.1 Generics tell the story

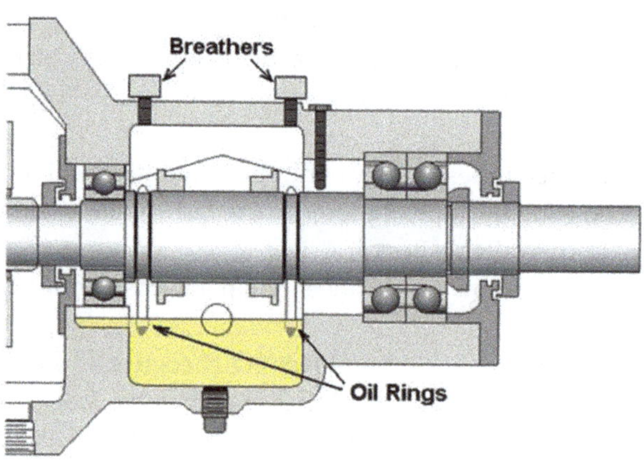

Fig. 2.14: A bearing housing with several flaws exposed.

The generic bearing housing cross-sectional view of Fig. 2.14 makes a compelling case for pump upgrading. It depicts at least five vulnerabilities and risks that are unacceptable to best-of-class users:

(a) Oil rings ("slinger rings") are used to lift oil from the sump into the bearings. Yet, under certain adverse conditions, these oil rings will malfunction [12]. Understand how and why they tend to malfunction [13].
(b) The back-to-back oriented thrust bearings are not located in a cartridge. However, cartridge mounting will facilitate upgrading from high-risk loose-running oil rings (slinger rings) to lower-risk flinger disks. Flinger disks are secured to the shaft.
(c) Bearing housing protector seals are missing from this picture. Not using bearing protectors or installing an inferior type of bearing protector will decrease bearing life. The incremental cost of a good product often pays for itself in a mere few weeks of ownership.
(d) While the bottom of the housing bore at the *radial* bearing shows the desired oil return passage, the same type of oil return or pressure equalization passage will be needed at the 6 o'clock position of the *thrust* bearing.
(e) There is uncertainty as to the type or style of constant level lubricator that will be provided; unless specified, the best one is rarely found on new pumps.

In addition to the more widely publicized hydraulic selection criteria, each of the above mechanical or power-end issues merits further explanation and will be discussed next. The considerations are confined to lubrication issues on process pumps with liquid oil-lubricated rolling element bearings. A majority of process pumps in use today fall into the oil-lubricated rolling element category. (Small pumps with grease-lubricated bearings and large pumps with sleeve bearings and circulating pressure-lube systems are not included in this discussion.)

2.5.2 Black oil

It has been estimated that 25% of all failures are preventable but not prevented because of an arbitrary decision that is simply not rooted in knowledge and experience. Example: A decision to use the cheap oil may overlook the fact that cheap oils often lack demulsifiers, or anti-foaming agents, etc.

Warding off "black oil" requires close observation of possible contributors. Black oil belongs to the approximately 15% of all failures that are predictable but not predicted. Example: The random appearance of "black oil" may be attributable to O-ring degradation of a certain style of bearing protector seal. The bearings will soon fail but few people have read the books and articles that describe the occurrence. (The occurrence could be linked to a certain risky design feature on a widely used product.) Or, as was shown in Fig. 2.11, oil could be trapped behind a bearing.

2.6 "FRETT" – force, reactive environment, time, temperature

Bearings, lubricants, and lubricant application methods all merit close attention. It is important to realize that, whenever gaskets, bearings, seals, shafts, housings, or any of the many thousands of other equipment components fail, the underlying causes are always found in one or two of four mechanisms described by the acronym "FRETT." This acronym stands for force, reactive environment, time, and temperature [7, 14]. So, whenever "black oil" is found in a bearing housing, at least one of these four failure mechanisms will have contributed to the seemingly random and often very sudden appearance of darkened lubricant.

While black oil has been studied in the past, the underlying causes seem to have remained hidden from many researchers [12]. An oil analysis will uncover contaminants that are either carbon (overheated oil) or particles of elastomeric O-rings that have frayed or decomposed for some reason. In any event, one will have to find and eliminate the source of the black residue. "Black oil" inevitably originates either from overheating the oil or is due to O-ring damage [12]. Both deserve much more attention.

2.6.1 Bearing housing protector shortcomings

Among the more elusive reasons for overheated oil, we find oil ring slippage and also degradation of a so-called dynamic O-ring that is placed between the rotating and stationary components of a bearing protector seal. Common sense and experience are of great value here; accordingly, a preview of bearing housing protector seals is enlightening:

Bearing protector seals with O-rings contacting a generously contoured area (Fig. 2.15c) are obviously less likely to undergo O-ring degradation than products designed with a sharp groove opposite the O-ring (Fig. 2.15a).

2.7 DN-number points to oil level preferences in bearing housings

2.7.1 Oil rings, general

Perhaps half of the world's oil-lubricated industrial pump bearing housings are furnished with inexpensive oil rings, also called "slinger rings," that lift the lubricant from an oil sump (Figs. 2.16 and 2.17). In some applications, oil rings or small disks that just barely contact the oil at its level surface are helpful for maintaining the oil volume more uniformly mixed, that is, to prevent temperature stratification. Stratification is a term that describes hot oil floating upward and staying near the top of the

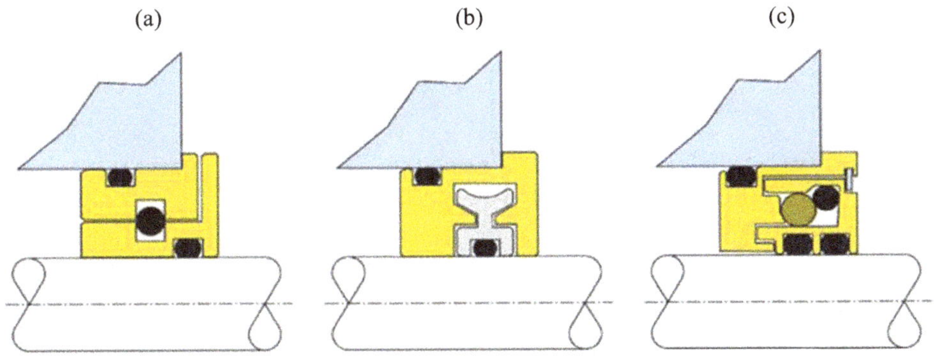

Fig. 2.15: Two generic bearing housing protector seals with configurations that risk O-ring degradation (left, "a") due to contact with sharp grooves, using a V-contoured ring (center, "b") that tends to increase frictional drag and, in contrast, modern design (right, "c") with optimally placed O-ring locations (source: AESSEAL Inc., Rotherham, UK; and Rockford, TN).

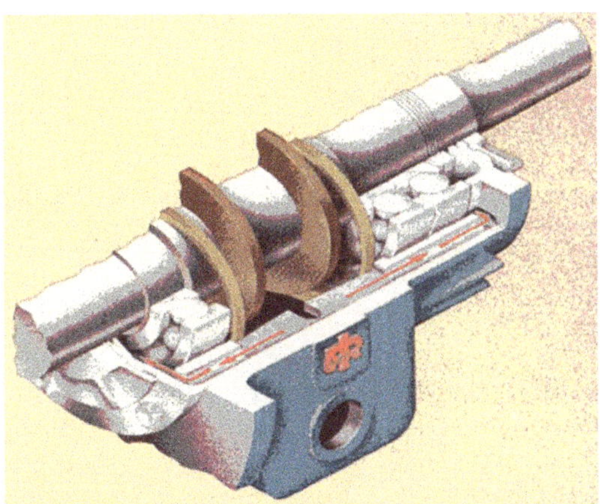

Fig. 2.16: Bearing housing with two oil rings ("slinger rings") and oil passages ("galleries") allowing oil to flow into the bearings.

oil sump. Conversely, the cooler oil has greater density and tends to sink down to the bottom of a bearing housing.

Typically, the loose oil rings shown in Figs. 2.16 and 2.17 are immersed about 3/8th inch (~10–11 mm) for the purpose of spraying oil into the bearings [15–17]. Oil rings are not required in pumps with low-to-moderate shaft peripheral velocities. In these typical "without oil ring" designs, the oil level must reach to the center of the lowermost bearing element – usually a bearing ball. Higher shaft velocities are defined next.

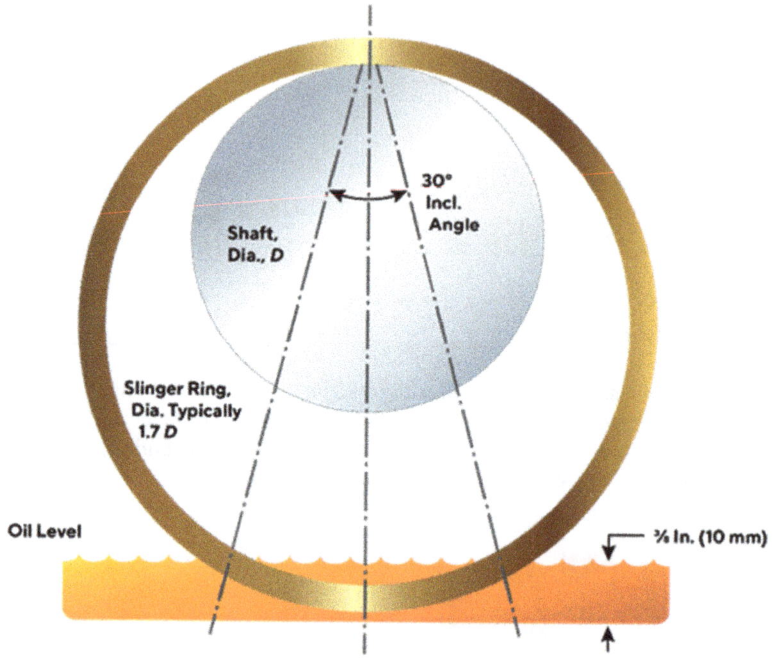

Fig. 2.17: Oil ring dimensional ratios and typical immersion depth (source: T. F. Hudgins, Houston, TX).

2.7.2 Bearing and shaft velocity constraints

Bearing peripheral velocities are more conveniently expressed as DN-values, the product of multiplying inches of shaft diameter times shaft rpm. For example, a 70 mm (~2.75″) bearing operating with a shaft turning at 1,800 rpm would have a DN-value of 4,960; values below 6,000 are somewhat arbitrarily considered low-to-moderate DN numbers. In designs with DN < 6,000 (conservative) and possibly reaching (but not exceeding) a DN as high as 8,000, both housing geometry and constant level lubricator height settings are generally selected to allow lubricating oil to reach the center of the lowermost bearing ball or rolling element. Keep the numbers 6,000–8,000 in mind; they represent an interesting DN parameter [15, 17].

Satisfactory oil ring functioning is obtained in the range of from 2,000 to 2,500 ft/min shaft surface velocity, which can be expressed in terms of diameter and rpm. Some bearing manufacturers (and also machine designers) recalculate to show the acceptable range of surface velocities comply in the metric system. Bearing producers often use a dN expression, with "d" generally designating a mean, or average, diameter (mm) of rolling element bearings. The mean, or average, dN would equate to [½] × [OD + ID] × rpm and we have seen dN-range stipulations from 300.000 to maximum values of 500,000. As an example, a bearing with a 100 mm bore, an OD of 180 mm, and a shaft operating at

3,000 rpm would have dN = ½ [100 + 180] × 3,000 = 420,000. The shaft surface velocity would be 100 × 3.14 × 3,000 = 942,000 mm/min or 3,090 ft/min.

Seeking reasonable agreement with the numbers given above, we looked up a 400 mm (~15.7″ shaft diameter) double-row spherical roller bearing used in a large mixer-agitator drive. With grease lubrication, the manufacturer allows a maximum speed of 380 rpm; its DN would thus be (15.7) × (380) = 6,000. With oil lubrication, the manufacturer advises a maximum speed of 480 rpm. Its DN would equal 7,540. The same bearing's outside diameter is 650 mmm, its metric "d" is (650 + 400)/2 = 1,050/2 = 525 mm. Therefore, dN for grease would be (525) × (380) = 199,500; dN for oil = (525) × 480 = 252,000.

Our conclusion would be that this shaft and its bearing system approach and possibly exceed the upper limits of shaft surface velocity. The equipment manufacturer tested the machine for a few hours at close to best achievable conditions of oil viscosity, shaft horizontality, ring immersion, etc. He acted in good faith when he shipped the machine to your plant. However, operating conditions at your plant are far from ideal, and the machine will likely experience an above-average number of failures. Perhaps it would have been prudent to upgrade the lubrication system at the specification stage, or to do so now, by identifying and scheduling a combined repair and upgrade task [13].

2.7.3 Oil levels in bearing housings with different size bearings

Reaching the center-of-rolling element oil level requirement would be difficult to achieve if the shaft support bearings had different diameters. In that case, one uses oil rings, as shown in Figs. 2.16 and 2.17. Oil rings are also used for higher speeds, in situations where oil reaching the center of the lowermost bearing is no longer reliable. Unreliable lubrication would result if the frictional energy of bearing elements racing through a pool of oil would be high enough to create excessive heat. However, serious concerns about the reliability risks associated with oil rings have prompted reliability professionals to seek more reliable means of lubricant application.

Some have elected to apply the lubricant by one of several possible pressure-developing pump-around means. Figure 2.18 shows one such circulating oil unit. Although originally designed for pressurizing the space between dual mechanical seals, the unit is well suited for bearing lubrication. Other thoughtful remedies and upgrades from oil rings include flinger disk retrofits (see index) or small oil mist units (Fig. 2.19). Experience checks are needed for small oil mist units and the manufacturer's operating procedures should be consulted.

Recall that with oil levels in bearing housings reaching the center of the lowermost bearing ball while allowing DN values *above* 8,000, potentially excessive friction-induced temperatures are likely resulting. Experiencing such temperatures

Fig. 2.18: Self-contained oil pump-around unit that can take over when and where oil rings fail (source: AESSEAL Inc.).

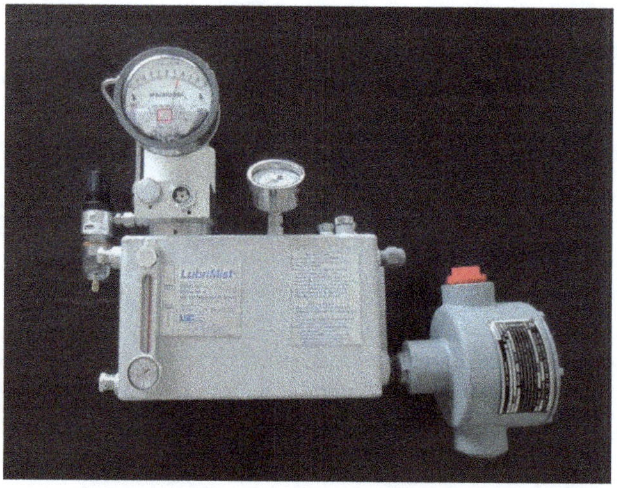

Fig. 2.19: Small self-contained oil mist unit (source: Lube Systems Company, Houston, TX).

led to the decision to lower the oil levels in many of the larger API-compliant pumps which operate at 3,000 and 3,600 rpm. In these, oil levels are customarily set well below the periphery of even the lowermost bearing ball. Because oil is then no longer flooding a portion of the lowermost part of the bearing, mechanical means must be employed to feed, lift, spray, or splash the lubricant into the pump bearings. Also, whenever high DN or dN values rule out letting lubricating oil reach the center of the lowermost bearing ball or rolling element, oil rings are often the (initially, but not ultimately) least expensive means of feeding oil into the bearings.

2.8 Oil rings have serious limitations

We have earlier alluded to oil rings (sometimes called slinger rings) having serious limitations, and some of these are either little known or not well publicized [14–17]. Oil ring instability ("wobbling") is not a new phenomenon. To ensure proper operation, surface velocity limits around 3,500 to 4,000 fpm (~18–20 m/s) *with* water cooling are often cited [17]. Such cooling simply implies the need for maintaining constant oil viscosity by closely controlling oil temperature. *Without* water cooling of the lubricant, we are asked to stay well inside the stable limit for oil rings and to not exceed peripheral velocities of 2,000 to 2,500 fpm, about 10–13 m/s [17].

Another source of information, a major multinational corporation's "Lube Marketing Course" text, suggests a DN value of 6,000 as the threshold of instability for oil rings. As a precautionary rule, both authoritative texts warn that oil rings in field situations tend to become unstable whenever DN, the product of shaft diameter (inches) and speed (rpm), enters the region from 6,000 to perhaps 8,000; we might allow 8,000 in a perfect test stand-like installation and assume that 6,000 is the limiting value for the realistic field installation. A 2″ shaft at 3,600 rpm would have a DN value of 7,200; it would thus operate in the risky or instability-prone zone. In comparison, equipment with a 3″ shaft operating at 1,800 rpm (DN = 5,400) might utilize oil rings without undue risk of instability [17, 18].

As just one more example, a nominally 3″ diameter (actual bearing bore = 75 mm) shaft at 3,600 rpm would have a DN = 10,800; it would definitely operate in the risky region. The surface velocity of this 75 mm shaft at 3,600 rpm would be $(\pi D/12)(3{,}600) =$ 2,780 fpm (~14.2 m/s). We must distinguish between situations with closely controlled (optimum) oil viscosity where the risky zone starts at 3,500 fpm, and less well-controlled oil viscosity and immersion, where the risky zone starts at 2,000 fpm [17].

2.8.1 More on test stand versus field experience

Earlier in this text we had mentioned that pump manufacturers will often point to satisfactory test stand experience at higher peripheral velocities than mentioned in the preceding paragraph. However, test stands operate under ideal conditions whereas field situations are usually far from ideal. In real-life or "field" situations, shaft horizontality and oil viscosity, depth of oil ring immersion, bore finish, and out-of-roundness are rarely perfect. Again: to be safe, we might opt to use either the DN < 6,000 or the surface velocity < 2,000 fpm "rule of thumb." Either way, the vendor's test stand experience is of academic interest, whereas the field experience that led to these rules of thumb should be given much weight. Reliability-focused plants are governed by well-documented field experience. They record this experience and pass it on to successive generations by a judicious combination of mentoring, training and, in some cases, collecting articles and mandating the reading of technical texts.

But think back of a situation where equipment with rolling element bearings was routinely supplied with oil rings. Assume the manufacturer is either unable or unwilling to offer superior lube application methods (such as undeniably represented by pure oil mist) in pumps with high DN-values. In those instances, polymer oil rings, as recommended by some pump manufacturers, will be a (very minor) step in the right direction. Still, it should be kept in mind that *all* oil rings have limitations that can be explained by experimentation, by direct observation (measurements), the groundbreaking texts of Wilcock and Booser [17] or by simple physics. Recall Fig. 2.12, which contrasted a new oil ring (left) and an abraded version (right). In [18] the two researchers reported on the need for oil ring roundness {concentricity) to stay within 0.002″ (0.05 mm). This requirement tells us that oil rings must be heat-stabilized and then machined with great care. Where these special manufacturing steps had been disregarded, the author measured out-of-roundness values of 0.017″ in case "A" and 0.062″ in case "B." Each failure event was accompanied by massive bearing failures and significant losses of revenue.

2.9 DN number concerns re-emphasized and summarized

Oil rings were probably first used on slow-speed machinery in the late eighteenth Century. Equipment speeds have since been increased and delivery methods re-evaluated. Here are some of the findings:

(a) oil rings ("slinger rings") are subject to a DN limitation; that is, certain rpm-times-oil ring-bore values should not be exceeded.
(b) oil rings are viscosity-sensitive and will perform properly in only a rather narrow range of lubricant viscosity.
(c) oil rings are immersion-sensitive and are affected by viscous drag. This drag is approximately proportional to the shaft surface velocity.
(d) oil rings must not be out-of-round or "slightly oval." Reliability-focused users should not allow more than 0.002″ (0.05 mm) ring eccentricity [18];
(e) oil rings will often run down-hill; they tend to malfunction if the total shaft system is not truly horizontal. In that case, they often contact stationary parts and abrade (Fig. 2.20)

(f) oil will get trapped in the space behind bearings unless a drain hole allows oil to return to the sump. Trapped oil will overheat and become coke.
(g) oil rings can get trapped in the clearance between the travel-limiting screw and shaft periphery (to the right of the oil ring closest to the thrust bearing in Fig. 2.14)

Overlooking (e) becomes an elusive reason for pump failures. Also, revert back to Fig. 2.1 and assume that the installer aligned the pump in its cold stand-still condition with the support leg under the thrust bearing firmly secured in place.

As this foot-mounted pump is later started and achieves its normal operating temperature, thermal expansion causes the casing to rise slightly while the support leg remains surrounded by cooler ambient air and will not thermally expand. The shaft system will then no longer be fully horizontal. An installer or millwright should use a procedure that calls for the support leg to be loose during pump alignment. It should be fastened after the pump has reached operating temperature [7]. In many instances the support leg can even be discarded because it really serves no purpose.

Fig. 2.20: Oil ring skew and/or downhill-running shafts will cause a number of problems (Source: Robert Matthews)

2.9.1 Shaft horizontality and oil level

Although industry now has access to, and benefits from, superb laser-type shaft alignment tools, an installer or millwright typically shims up one end of the pump, thereby jeopardizing shaft horizontality [19]. Also, the user will experience occasions involving lubricant delivery via “constant level” lubricators, the true oil level will actually be lower than the set point indicated on the constant level lubricator assembly. Understanding how these lubricators function is very important and will save much pain.

2.10 Constant level lubricators

The most typical application of oil involves using one of many available constant level lubricators. A widely used version is shown on both sides of the bearing housing in Fig. 2.21. However, side-mounted constant level lubricators or oilers are unidirectional. For proper operation, a constant level lubricator ("oiler") should be mounted on the up-arrow side of the bearing housing. Because the shaft rotation in Fig. 2.21 is given as counterclockwise the oiler shown on the left side should be removed to reduce the risk of air being aspirated or pulled in. The constant level lubricator shown on the right side should stay; note that it is mounted on the "up-arrow" side of the counter-clockwise rotating shaft and surrounding bearing housing system.

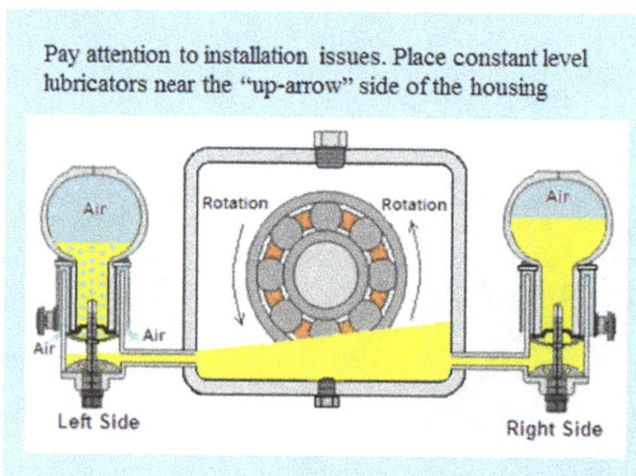

Fig. 2.21: Traditional liquid oil application with static sump. The lubricator on left side should be removed; non-balanced lubricators are shown (source: Trico Mfg. Corporation, Pewaukee, WI).

Unless the lubricator is properly mounted on the up-arrow side of the shaft's rotation, there will be an increased risk of the oil level lowering, suddenly depriving a bearing of lubrication [7, 10].

Also, if the pressure in a closed bearing housing increases due to a slight temperature increase, the resulting pressure increase will cause the oil level to go down and the oil may suddenly no longer flow into the bearing [20]. The top layer of oil in the sump will quickly overheat and black oil will form. At that point, the bearing will begin to fail. Pressure balancing lubricators (see Figs. 2.22 and 2.23, right side) are much preferred over unbalanced types.

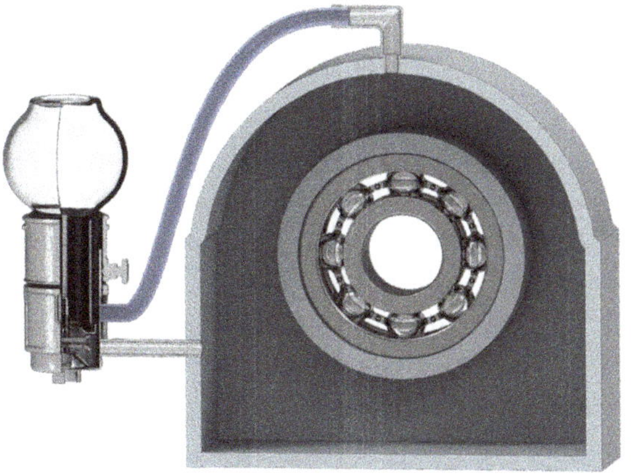

Fig. 2.22: Typical pressure-balanced constant level lubricator mounted on clockwise-rotating unit.

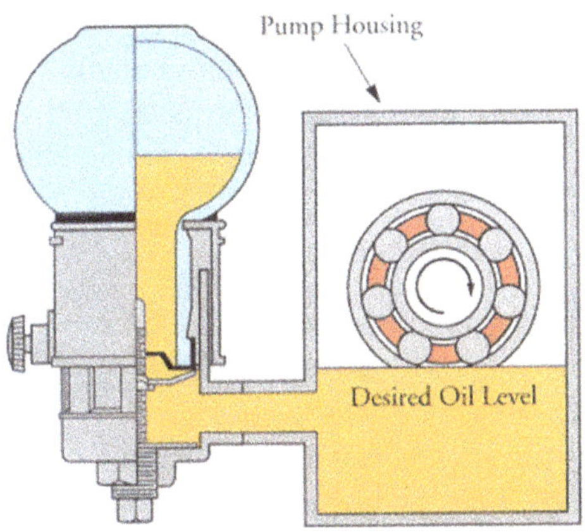

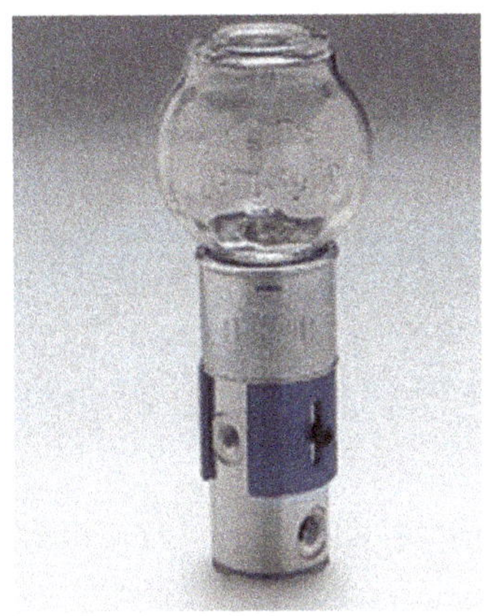

Fig. 2.23: Non-pressure-balanced constant level lubricator (left) and pressure-balanced version (right, shown here without sight glass. Note balance line port) (source: TRICO Mfg. Corp.).

Caulking was applied by the oiler manufacturer where the transparent oiler bulbs in Figs. 2.20–2.23 join their cast metal support bases. Since this caulking is subjected to numerous repeat swings in ambient temperatures, the caulking has a finite life. Rainwater runs down on the glass and meets fissures or microcracks in the caulking. Capillary action pulls the water into the microcracks and reaches the lube oil in the cast metal support frame. Water contamination of the oil is always detrimental to bearing reliability and, as a consequence, oilers should be replaced every few years. Sadly, some commercial entities are among the uninformed people who benefit from low reliability and stand to profit from the uninformed user requiring frequent repairs.

2.10.1 Making informed choices

When all is said and done, engineers must make a choice. That choice is to either follow the indifferent pack or follow the true professionals who treasure the truth. Our advice for engineers and reliability technicians is to study the facts, understand the science of lubrication (tribology) and then teach others. The truths must never contradict science and must always harmonize with common sense [21].

Jumping ahead and referring to oil mist lubrication, suppose you encounter resistance when you advocate oil mist for your hot service pumps and electric motor drivers. Place books and articles in the hands of the uninformed staffers who, perhaps quite often, convey flawed or erroneous information. Ask them to do some research and come back in a few days with an answer to the question why pure oil mist works beautifully on hundreds of hot service pumps at Competitor "X" but not here, at your Plant "Y." Reason with ones who tend to hold back progress. If reliability and profitability-improving lubrication works well at the close-by oil refinery "X" but does not work well at your oil refinery "Y," you must find out and isolate what people at "X" are doing differently. In other words, and to emphasize without equivocation: If oil mist works well on hundreds of seemingly identical pumps elsewhere, the problem where it does not work must be with the employees or managers at the plant. If oil mist is successfully used in 52,000 electric motors and your staffers claim that oil mist systems cannot provide lubrication for the motors at your location, it is time to ask some probing questions and demand factual answers.

2.10.2 Why use only pressure-balanced constant level lubricators

There have been many instances where the transparent reservoir bulbs of constant level lubricator assemblies showed adequate levels of oil which, understandably, led operating technicians to *assume* the oil level in the bearing housing was at the correct height. Examining simple old hydraulic laws should convince us that such an assumption is not always correct. The lube oil level in a bearing housing cannot coincide with the level in the base of the constant level lubricator if the internal bearing housing pressure is different from atmospheric pressure. It should again be noted that, in a typical *non-balanced* constant level lubricator base (Fig. 2.21), the oil level is contacted by ambient air. According to the laws of physics, slightly elevated pressures in the bearing housing will drive the oil level lower in the bearing housing. In bearings and housings where oil levels reach the center of the lowermost ball, lowering the oil level by a few thousandths of an inch may be the difference between oil flowing into the bearing and no oil flowing into the bearing. A bearing without oil will soon overheat and the oil will turn black [22].

If the pressure in an inadequately vented bearing housing rises above ambient pressure, the displaced volume of oil will rise in the metal base of the constant level lubricator. As a not-to-be-neglected separate but additional consideration, ambient air in most industrial locations carries entrained water vapor and airborne contaminants. These can enter the non-balanced constant level lubricator below the transparent bowl and cause, ultimately, a decrease in bearing life.

Properly *pressure-balanced* constant level oiler assemblies are recognizable by a balance line connecting the pump bearing housing to the location where the transparent bowl meets the oil level in the constant level lubricator (Fig. 2.22).

The resulting pressure equalization together with a suitable bearing protector seal often cures bearing lubrication problems. A fully enclosed oil-lubricated bearing environment is made feasible. Compared with non-balanced constant level models, the typical incremental cost of an average-size pressure-balanced constant level lubricator with sight glass option (Fig. 2.22) is perhaps close to \$40–\$60 (in 2022). If 200 pumps were being retrofitted with advanced bearing protector seals and pressure-balanced lubricators, the value of the resulting failure avoidance for even a single unscheduled pump downtime event in each of the next 5 years would make overwhelming economic sense. The peace of mind procured with an incremental outlay of perhaps \$60 or even \$100 is hard to quantify; still, such routine upgrading whenever pumps are being repaired or serviced would represent tangible and verifiable reliability improvement steps for process pumps. Updating the plant's procurement specifications and initially purchasing balanced constant level lubricators for machines that for some reason are sold with constant level lubricators would be even better.

2.10.3 Disseminating information relating to lubricators

From an operating and maintenance perspective, one should not overlook that there must always be a partial vacuum at the top of constant level lubricator bulbs. This is the kind of detail that must be part of refresher training for operators. At best-in-class companies, refresher training is tacked on to the obligatory safety or toolbox sessions. One corporation calls these 5-to-7-min add-ons "Shirt Sleeve Seminars" because the presenters may have to roll up their sleeves to deliver these messages [14]. At the conclusion of each such safety session and added shirt sleeve seminar he or she hands out a three-hole punched two-sided laminated-in-plastic sheet with important reliability-enhancing reminders. Mechanics and machinists were given three-ring binders and are asked to place these reminder sheets in their own personal three-ring binder.

Although operations control rooms have binders containing operating procedures and perhaps a few operations-related asset preservation guidelines, more is needed. Operating managers are expected to ascertain that, in addition to up-to-date plant operational procedures that can be viewed on computer screens, relevant handouts are placed in three-ring binders at each control room.

By way of just one more reminder that should be of interest to different job functions throughout a company: The various constant level lubricators shown earlier cannot possibly function if they have been filled to the top. Also, if caulking is used to cement a transparent bulb to a metal body, it must be realized that this caulking has a finite life. Once it develops tiny aging cracks, rainwater may enter by way of capillary action. Finally, traditional constant lubricator assemblies are direction-sensitive and should be mounted on the "up-arrow" side of the bearing housing. This was indicated on the shaft in Fig. 2.21 and both the manufacturer's instruction manuals and other texts certainly describe this requirement.

2.11 Needed: a better choice than oil rings and constant level lubricators

As was observed earlier, most 3,000 and 3,600 rpm pumps obtain splash lubrication through the action of oil rings (also called slinger rings). Because oil rings have inherent shortcomings that often make them a poor choice for risk-averse plants, flinger *disks* (Fig. 2.24) securely mounted on the shaft are often preferred [15]. Properly designed flinger disks offer much lower risk of malfunctioning than oil rings. Many reputable European pump manufacturers use flinger disks, and some have supplied them on their pumps for many decades. Since flinger disks are secured to the shaft, they are less affected by shaft horizontality, oil viscosity, immersion depth, shaft surface finish, and ring concentricity. These five factors inevitably vary from pump to pump; they result in an infinite combination of variables and some of these combinations tend to make oil rings rather prone to malfunction.

An earlier description pointed to flinger disks (Fig. 2.24). They are securely fastened to the shaft and avoid the slippage and abrasion problems encountered with loose oil rings. However, whenever metal flinger disks are manufactured with diameters larger than the bearing housing bore, a bearing may have to be mounted in a cartridge. The outside diameter of this cartridge must be large enough to allow passage of a shaft with a flinger disk secured to it. Because these mounting cartridges are precision-made and will add to pump cost, they are not found in all pump models.

In any event, the vagaries of constant level lubricators and observation of relatively sudden appearance of "black oil" illustrate that the basic laws of physics apply even to seemingly insignificant parts of machinery. That said, pump operators and reliability professionals need to reflect on these issues and understand every one of the various risk factors. If the remedies offered by a pump manufacturer resemble Fig. 2.25, ask questions before you accept the proposed solution [16]. Perhaps the design was used on shipboard where wave motion makes loose oil rings impractical, to say the very least.

But the proposed design shows an oil ring with sharp edges on both sides. As these sides contact the tall slightly tapered vertical inside boundaries of the carrier spool, will they not tend to break through the oil film coating and will there not be metal-to-metal contact? Abraded particles would contaminate the oil and bearing life would be shortened,

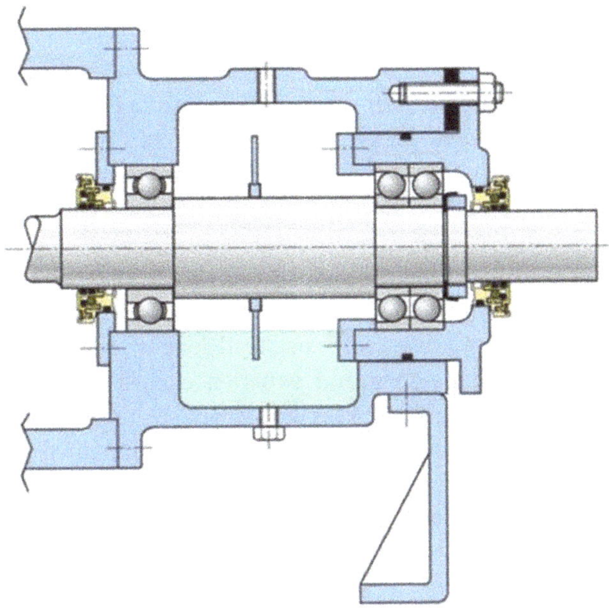

Fig. 2.24: Shaft-mounted flinger disk and bearing cartridge at thrust end.

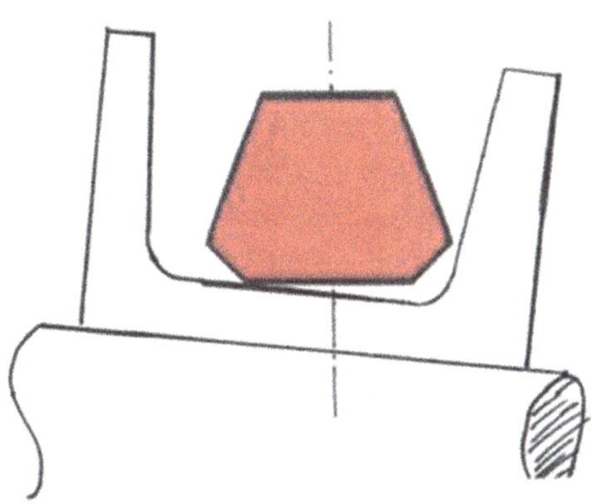

Fig. 2.25: Loose oil ring running in a ring carrier (source: Simon Bradshaw, Proceedings of 17th Texas A&M University International Pump Users Symposium, Houston, TX, 2000).

Pump owners must accept that, as different risk factors combine, one single additional small deviation could bring down the entire plant. Therefore, it should be a priority to examine all probable causes and factors that so often combine and contribute to costly repeat failures of pumps. It follows that judicious upgrading should be pursued. Also, the findings of [17, 18] are very relevant and will have been heeded by competent pump rebuild shops.

2.11.1 Rebuilding and upgrading are urgently needed

Next time a process pump fails, a particular owner-operator is handed a fine opportunity go beyond the typical repair. It would be appropriate to upgrade the pump, and to thus greatly reduce the probability of repeat failures. But first, a competent pump repair shop or pump rebuilder must be selected. So, why should that be an issue worth mentioning?

As is so often the case, there are pump repair and rebuild shops whose entire focus is on keeping cost low. These shops are often unable to provide the engineered solutions needed for best efficiency, lowest life cycle cost, and longest equipment life. Performing both an up-to-date competency and experience check together with understanding and applying the concept of life cycle cost will steer the owner-operator to the right rebuilder. This rebuilder will be ready and willing to prove a solid combination of planning and work execution expertise.

The "rebuilder of choice" must perform and report to the owner-operator a large number of critical measurements. Also, the rebuilder must have a proven track record and assist the owner-operator of the process pump in compiling a list of deliverables. Pump owner and pump rebuilder must cooperate in performing a structured audit aimed at developing a written list that describes and confirms the adequacy of the rebuilder's detailed procedures. The audit includes ascertaining the availability of a rebuilder's experienced work force to carry out the projected upgrade work [19].

The shop's needed degree of solid pump experience is best described by the following example, dealing with lubrication compromises that must be addressed. Pump manufacturers and rebuilders should be aware of special issues and potential problems that arise if, in any pump, both sleeve type and rolling element bearings share the same bearing housing. In essence, viscosity is of greatest importance and each bearing type fails if there is prolonged metal-to-metal contact.

To prevent contact, the oil film in sleeve bearings has to be thicker than the asperities (the surface roughness) in the bearing and journal surfaces. The oil film in a sleeve bearing must also be thicker than an occasional dirt particle traveling in the oil. Most importantly, the oil needs to be properly applied and must form a suitable film on the surfaces where such a film is most needed.

2.11.2 Test the pump rebuild shop's lubrication knowledge

To make a long story short, a lubricant with all required performance attributes, including film strength and film thickness must also allow oil rings to function properly. Only a well-proven <u>*synthetic*</u> ISO Grade 32 will satisfy most requirements. Preferred suppliers usually formulate such lubricants from a PAO/diester synthetic base oil to which an ionic bonding agent has sometimes been added. The value of superior lubricants and access to competent providers with application engineering knowledge far exceeds its incremental cost over least-expensive oils.

For best results and highly satisfactory long-term performance, the oil will have to be a premium synthetic ISO VG 32 or VG 46 formulation. It must give users the protection and film thickness/film strength properties of ISO VG 68 mineral oils. While these properties may not be needed elsewhere in one's facility, they will make lots of technical sense in many process pumps that use both sleeve bearings and rolling element bearings in the same bearing housing [20]. Find out if the pump rebuilder is aware of this issue.

During troubleshooting lubrication issues on existing pumps, one will often find sludge in pump bearing housings. It should be realized that water acts as a catalyst that promotes sludge formation. Sludge is thus often the result of water and atmospheric dirt, in addition to oil ring (slinger ring) debris. Exposure to airborne particulates is unavoidable in some environments and water intrusion is possible in other environments. Another potential source of water is from cracked water jackets. There is some irony in that observation, since cooling water may not be needed in the first place [21]. Again, be sure the pump rebuilder understands this fact and supports your decision to delete cooling.

2.12 Why avoid low-cost lubricants and lube delivery methods

Pump owner-operators should become familiar with the inadvisability of mixing two "virtually identical" oils from different suppliers. There is ample evidence that, so as to keep costs low, some oil suppliers skimp on the amount and/or quality of additives. The lowest-cost-supplier game is quite often encouraged by users. It stands to reason that engaging in these ill-advised procurement practices will result in suppliers and users not attracting, grooming and/or retaining top talent.

If the user plays along with the supplier's low-price strategy, the latter feels encouraged to provide a "commodity product" and commodity lubrication products may not serve those of us who go beyond paying lip service to the term "reliability." Are your lubricants selected on pricing criteria alone? Did someone in authority decide to "standardize" on one lubricant type for the entire facility? If you answered these two questions in the affirmative, your plant reliability is trending downhill toward below-average reliability and profitability.

An earlier comment recounted an oil ring that had been removed from a large process pump and was measured after one more of many repeat pump failures. Instead of staying within the allowable eccentricity of 0.002," this ring was found about 0.06" eccentric – 30 times the allowable value.

The ring had been abraded on one side; it also showed very serious discoloration – similar to Fig. 2.12. Reliability professionals should question the risks with oil rings (slinger rings). If slinger rings must really be utilized, they should never be the cheap variety. Good slinger rings will have gone through an annealing step before finish-machining. Again, reliability professionals must subject slinger rings to rigorous specifications and quality control. Width and concentricity of these oil rings (slinger rings) should be measured before the mechanic or machinist installs them in a bearing housing, and also after a machine is dismantled for repair, perhaps years later.

On pumps furnished with rolling element bearings, the purchaser may be able to avoid oil rings/slinger rings altogether by insisting on bearing housings configured to accept solid flinger disks. If nothing else, specifying a lube application method other than oil rings/slinger rings will start a discourse with the pump manufacturer. It should also be realized that oil rings/slinger rings can jump around when the equipment is transported from a factory or shop to an installation site. There have been occasions when slinger rings were later found lodged between shaft and ring locator pin or some other housing component; see earlier pages.

The point of the story is that truly reliability-focused purchasers are justifiably concerned with certain traditional and all too often failure-prone means of lube delivery. Pump manufacturers must be shown the vulnerabilities of certain pump designs and pump users must use rigorous checklists that lead to better installation procedures, verification of adequacy before startup, adequacy while running, and checklists while performing root cause failure analysis. Thousands of repeat failures occur in industry every day; they are irrefutable testimony to the fact that things are far from acceptable and that striving for improvement is a shared obligation that cannot be shrugged off by any of the parties involved.

2.12.1 Experience-based rankings for general guidance

In the early 1980s, a major bearing manufacturer with roots in Germany published a ranking list for industry. Several widely used oil and grease application methods were listed in the bearing manufacturer's order of preference. Years later and with the advent of improved lubricants, the author reshuffled these rankings. A provision was added to link the 1982-vintage recommendations (Tab. 2.1) to process pumps and their electric motor drivers only. The new rankings were compiled in 2014 and the left column of Tab. 2.2 should be assumed to reflect oil-lubricated pump bearings. The right column is based on the author's decades of experience with grease-lubricated electric motor bearings.

2.13 Understanding elusive bearing lubrication issues

Because rolling element bearings are standard components in most centrifugal pumps, scores of authors have devoted time and effort to subjects affecting bearing performance. Lubricant degradation is among the many topics, although not many articles have dealt extensively with oil ring-related contamination and "black oil" for-

Tab. 2.1: Generalized Lube Application Rankings Used by a Bearing Manufacturer in 1982.

Decreasing service life **	Oil	Grease
↓	Rolling bearing alone	Rolling bearing alone
	Circulation with filter, automatic oiler	Automatic feed
	Oil-air	
	Oil-mist	
	Circulation without filter*	
	Sump, regular renewal	Regular regreasing of cleaned bearing
		Regular grease replenishment
	Sump, occasional renewal	
		Occasional renewal
		Occasional replenishment
		Lubrication for-life

* By feed cones, bevel wheels, asymmetric rolling bearings.
** Condition: Lubricant service life < Fatigue life.
Source: Eschmann, Hasbargen and Weigand, *Ball and Roller Bearings*, John Wiley & Sons, Ltd., New York, 1985.

Tab. 2.2: Oil lubrication and grease lubrication, (author's experience-based rankings for pumps.).

New: Experience-based new ranking proposal, 2016

Oil (on pump bearings)	< <> >	Grease (on motor bearings)	
• Oil spray, filtered/pressurized Oil mist (globules of liquid oil	= 10		
• Circulation, filtered/pressurized	= 9	• Life-time PFPE-PTFE	= 7
• Quiescent sump, (debris settled)	= 6	• Through-flow, low pressure	= 6
• Sump, with slinger disc	= 5		
• Sump and guided oil rings, well within concentricity spec	= 4	• Random-flow, low pressure	= 4
• Sump and non-guided oil rings well within conc. spec	= 3	• Life-time EM polyurea	= 3
• Sump and non-guided oil rings out-of concentricity	= 2	• Random-flow, random press.	= 2
		• Random-flow, over-pressured	= 1

Key: 10 = Best; 1 = Least Confidence

mation over short periods of time. Needless to say, these are the "elusive" lubrication issues, and these are often involved in repeated short-term degradation of the lube oil. Equipment outages result and time and money will inevitably be spent. That's why the subject of "black oil" should be important to us. But as we will see, black oil formation often ties in with lubricant application methods. We must really understand the pros and cons, the misunderstandings and questionable claims made for some of the many different lube application methods.

2.13.1 Bearing housings with or without oil rings

By the author's estimate, about 45% of oil-lubricated pump bearing housings are furnished with oil rings dipping into an oil sump, as shown earlier in Fig. 2.17. Perhaps another 45% of oil-lubricated pumps are designed for operation without oil rings. In the typical no-oil-ring design the oil level must reach to the center of the lowermost ball, roller, or other bearing element. This level location requirement would be difficult to achieve if the bearings supporting the shaft were to have different diameters. Also, at higher rolling element speeds, plowing through an oil bath may generate too much heat, prompting the equipment manufacturer to lower the oil level and decide in favor of one or two oil rings (Fig. 2.16).

In the estimated remaining 10% of oil-lubricated pump bearings the lubricant is probably applied by pump-around ("pressurized") means shown earlier in Fig. 2.18, or the housings incorporate flinger disks, or pure oil mist is being used. In some applications, oil rings are also needed for the purpose of keeping the oil volume more uniformly mixed, i.e., to prevent stratification. The term "stratification" describes hot oil floating upward and staying near the top of the oil sump. Guidelines and application ranges for oil rings were discussed earlier in this chapter and are briefly re-stated here for emphasis.

2.13.2 Attempts to improve on troublesome oil ring lubrication

To re-emphasize: Oil ring instability ("wobbling") is not a new phenomenon. To ensure proper operation, [17] cites surface velocity limits around 3,500–4,000 fpm (~18–20 m/s) *with* water cooling. Stipulating water cooling implies the need for maintaining constant viscosity by closely controlling oil temperature. *Without* water cooling of the lubricant [17], advises staying well inside the stable limit for oil rings and not to exceed shaft peripheral velocities of 2,000–2,500 fpm (~10–13 m/s). Recall that a major multi-national corporation's proprietary "Lube Marketing Course" text, suggests using a DN value of 6,000 as the threshold of instability for oil rings. Recall also that both of these authoritative texts warn that oil rings in field situations tend to become unstable whenever DN enters the region from 6,000 and higher, see page 29.

We had explained on page 29 that a 2″ shaft at 3,600 rpm would thus operate in the risky or instability-prone zone, whereas equipment with a 3″ shaft operating at 1,800 rpm (DN = 5,400) might use oil rings without undue risk of instability. As just one more example, a 3″ (75 mm) diameter shaft at 3,600 rpm would operate with a shaft peripheral velocity of (π D/12)(3,600) = 2,827 fpm (~14.4 m/s). The fact that a pump manufacturer can point to satisfactory test stand experience at higher peripheral velocities was duly acknowledged, but field situations represent the "real world" where shaft horizontality and oil viscosity, depth of oil ring immersion, bore finish and out-of-roundness are rarely perfect. We can thus opt for using either the DN ≤ 6,000 or the surface velocity ≤ 2,000 fpm (10 m/s)"rule of thumb." A vendor's test stand experience is of academic value and the field experience that led to these rules of thumb should take precedence here. Avoiding oil rings is beneficial. When oil mist is considered, oil ring elimination should reflect in the cost justification for oil mist [21].

2.14 Black oil and bearing protector seals

There have also been reported instances of black oil formation at higher speeds and/or while using bearing protector seals. In one particular style of bearing housing protector seal with a dynamic O-ring contacting the sharp edges of a stationary groove, O-ring degradation (especially prevalent at slow-roll or turbine warm-up speeds) has caused lube oil contamination. It is not known if this eventuality was assessed in the research studies of the pump manufacturer making the black oil presentation in 2000. Also, it must be recognized that wide bearing protector seals designed with only a single O-ring clamping the rotating part to a shaft are inherently less stable than bearing protector seals that use two O-rings for clamping the rotating component to the shaft. At higher speeds a wide rotating component utilizing only a single clamping O-ring tends to align itself out-of-perpendicular relative to the shaft centerline. Whenever this happens, a dynamic O-ring in contact with a sharp O-ring groove will create havoc. Not only will it then drag excessively, but it may actually seize.

2.14.1 The story of "black oil" in pump bearing housings

After industry received many reports of "black oil," one major pump manufacturer decided to look into the matter of oil ring instability and the random events of black oil that had been experienced over the years. The resulting *"Investigations into the Contamination of Lubricating Oil in Rolling Element Pump Bearing Assemblies"* were published in the early 2000s and later again explained in a lecture in 2014 [16, 22].

The scope of both report and lecture was to give an analysis of factors affecting the short-term contamination of liquid lubricants in ring-oil lubricated rolling element bearings. "Short-term" was defined as intervals ranging from one hour to several weeks. The analysis dealt with the pump manufacturer's standard range of centrifugal pumps, and short-term oil degradation had been reported in several installations. However, although the pump manufacturer's original report was compiled in 1999, some large petrochemical corporations reported they were still struggling with the black oil issue in 2008.

2.14.2 More on the issue of darkened oil

We know from experience that closely observing both new and "used" oil rings will prove revealing. Abrasive wear of oil rings is easily recognized; a previously chamfered edge is now razor-sharp or burred (Fig. 2.12), or an originally straw-colored lubricant has recently turned dark. Debris from many copper-containing alloys leaves a grayish color; pure overheating generally produces floating carbon particles that cause the oil to turn black. One could also measure the "as installed" oil ring width (its thickness in the axial direction) and later compare it with its "as-found" width. Measuring will avoid the cost of analyzing for contaminant composition. Needless to say, flaked-off oil ring material is suspended in the oil and bearing life is cut short.

It was noted, in 2000, that this pump manufacturer's prior practice of using low viscosity "thin" ISO Grade 32 oil, while acceptable for "plain" or sleeve bearings, was risky for rolling element bearings. In many of the manufacturer's small and mid-sized pumps with rolling element bearings, thin oil simply exacerbated the problem of premature bearing failures. Obviously, to use a much more viscous "thick" oil would tend to slow down an oil ring. However, the pump manufacturer wrote about what was thought the solution to the problem and his report now recommended (in 2014) switching to another oil viscosity, ISO Grade 46 [22].

Again, we look at actual field experience. While oil viscosity is a very important parameter, tweaking the oil viscosity selection and substituting ISO Grade 46 (mineral oil) for the previously used ISO Grade 32 will not make much difference. Relatively minor variations in ambient temperatures negate the effect of small viscosity changes; many temperature- vs. viscosity charts included in bearing texts confirm this observation. Moreover, SKF had determined, well before 2000, that film thickness and film strength limitations rendered ISO Grade 32 mineral oils unsuitable for rolling element bearings in many centrifugal pumps. In its centrifugal pump handbook [20] SKF had asked users to restrict mineral oil ISO Grade 46 lubricants to bearings operating at temperatures not exceeding 70 °C (158 °F) and recommended ISO Grade 68 (again assuming mineral oil) for bearing operating temperatures not exceeding 80 °C (176 °F).

In slower speed bearings that are *not* using oil rings, black oil formation usually happens after the oil level has dropped by a mere fraction of an inch, essentially low enough to just barely reach (in horizontally arranged shaft systems) the bearing's outer ring shoulder. The outer ring shoulder is, occasionally, described as the edge (at the 6 o'clock position) of the bearing outer ring bore.

2.14.3 Oil level and oil application concerns must be addressed

There have been many instances where the reservoir bulbs of constant level lubricator assemblies showed adequate levels of oil which, understandably, led technician-operators to assume the oil level in the bearing housing was at the correct height. Hydraulic laws should convince us that such an assumption is not always correct. The lube oil level in a bearing housing cannot coincide with the level in the base of the constant level lubricator if the internal bearing housing pressure differs from atmospheric pressure. It should be noted that, in a typical *non-balanced* constant level lubricator base, the oil level is always contacted by ambient air. Obeying the laws of physics, slightly elevated pressures in the bearing housing will drive the oil level lower. Moreover, ambient air in most industrial locations carries water vapor and airborne contaminants, either of which will drastically decrease bearing life [23]. Groundbreaking research by Felsen, Cantley, and others is contained in [24–26]; it demonstrated how bad water is for bearings.

To recap, allowing a slightly higher pressure on one side of a bearing vs. the other side of the same bearing will cause the oil level on one side to vary from the level on the other side. Attempts to equalize pressure are made by milling a small slot (perhaps ¼″ or ~6 mm wide and 1/8th inch or ~3 mm deep) into the lowermost location of the housing bore. Without such a passage the expectation of pressures being equalized may be thwarted, especially if the bearing(s) are so-called angular contact "A/C" types. In some A/C bearings the slanted bearing cage (ball separator) inclination creates windage (or fan effects) that promote unequal oil levels.

The value of advanced bearing protector seals and pressure-balanced lubricators is easy to assess. Calculate the value of avoiding even a single unscheduled pump downtime event in each of the next 5 years and imagine the peace of mind those incremental dollars would buy. That would be a tangible and verifiable reliability improvement step for thousands of pumps. Routinely retrofitting pumps with these upgrade components can be readily accomplished during planned shutdowns or unscheduled pump repair events.

Referring again to constant level lubricators and from an operating and maintenance perspective, don't overlook that there must always be a partial vacuum in the upper part of the lubricator bulb. Some constant level lubricators may refuse to function as intended if filled to the very top, as shown in Fig. 2.26.

2.15 Needed: a better choice than oil rings

Most 3,000 and 3,600 rpm pumps obtain splash lubrication through the action of oil rings. Yet, with oil rings having inherent shortcomings that often make them a poor choice for risk-averse plants, flinger disks securely mounted on the shaft seem to offer a better choice than oil rings. Many reputable European pump manufacturers avoid oil rings and use metal flinger disks instead; some have done so for many decades.

Fig. 2.26: Constant level lubricators in overfilled condition cannot work as intended.

Since flinger disks are secured to the shaft, their performance is not subject to the influence of shaft horizontality, oil viscosity, depth of immersion, shaft surface finish, and ring concentricity. These five factors inevitably vary from pump to pump; they result in an infinite combination of variables and some of these combinations tend to make oil rings prone to random malfunction.

The author is familiar with years of efforts to "tweak" oil rings (see [16, 22]) and firmly believes that many of the findings are proof that deviations and extrapolations from the research findings in [17, 18] have little merit. Accordingly, a fundamentally new approach will be needed to be worthy of reliability-focused thinking.

2.15.1 Contemplating ideal lube applications

Some day in the future an enterprising inventor or pump manufacturer may turn his back on cost cutting and may surprise us with ingenuity. He may then develop a better alternative to oil rings or even flinger disks for rolling element bearings. It may be a smart device just short of the well-proven oil mist, and less costly than the widely known and often necessary pressurized oil pump-around systems. Perhaps reconfiguring the shaft to serve as the rotor of a progressive cavity pump or utilizing the principle of magnetic coupling might lead to a housing-internal means of picking up some oil, pressurizing a small stream of this oil, and then spraying it into the bearing rolling elements. There are also small gear pumps that could be placed in the bearing housing sump; a small gear pump could be driven by skewed gears on pump shafts. The possibilities are endless.

Meanwhile, industry must concentrate on proven lube application methods that have been used with great success, for decades and with costs well understood. Only oil mist matches these criteria [27]. Earlier in this text (Tab. 2.1), the ranking order of different lube application was explained. Oil mist was at the top of the list.

But is oil mist really cost-justified? If so, why do so few European oil refineries use plant-wide oil mist systems? These questions were addressed in an article published in 1990, where the author noted "together with an appropriate amount of a suitable state-of-the-art synthetic lubricant, this low-cost retrofit (referring to a modern magnetic seal and a plugged vent instead of the customarily open-to-atmosphere bearing housing vent port] may extend bearing life to the point where oil mist lubrication is no longer economically attractive" [21].

This statement is probably as true today (2021) as it was in 1990. But the context is of great importance, because it (i.e., [21]) referred to the small but diligent group of equipment users that insist on correct pump installation, operation, and maintenance. For them, it may indeed be difficult to cost-justify oil mist lubrication. These are the relatively few facilities that expertly apply the right lubricant to a particular bearing and change the oil periodically – sometimes as often as four times a year! Using a highly trained workforce and holding them fully accountable, industry in many European countries apparently does not experience enough bearing failures to justify additional (incremental) failure avoidance through the use of plant-wide oil mist systems.

There are many elements that contribute to this remarkable difference. The European mindset appears more oriented toward taking the necessary time to do things right whereas, on this side of the Atlantic, the mere speed with which a repair is carried out is often given more weight. In Europe, the administrative person in charge will not (usually) interfere with the experience-based judgment of a highly qualified craftsperson. If periodic oil changes are needed, they will be performed. If bearing installation tools are needed, these will be procured and properly used. Piping will be installed with proper fits, pulling pipes into place is considered a violation of the craftsman's "do no harm" ethic.

Regrettably, the European approach is rarely practiced in some parts of the world. All too often the persons in charge may insist on quick work; they will not allocate the time it takes to understand and remedy the underlying causes of failure. When our typical person in charge manages to quickly restore equipment to running condition, he or she will be celebrated and elevated. If the quick fix attempt fails, blame for having guessed wrong can usually be shifted to others. Once we get away with a deviation, deviations from the original quality norm become the new norm, and costly repeat failures are experienced [28]. Be forewarned, though: Repeat failures are the precursors of extreme failures. When *extreme failures* occur on lubricated equipment, there is usually a fire [7].

Experience shows that virtually all US facilities will likely benefit from a combination of judicious upgrading and the application of pure oil mist lubrication in a closed loop [29, 30]. Ample documentation and references will be found in Chapter 3.

2.15.2 Accountability

Whenever truly pertinent training and accountability are lacking, the cycle repeats itself. As just one example, in many plants no one can explain why and how a constant level lubricator works and why the widely used non-balanced versions no longer represent state-of-art accessories. Lube replenishing duties are often overlooked or carried out wrongly. Inadequate oil rings continue to be used in many thousands of bearing housings, and defensive bickering is often preferred over sitting down and listening to solidly science and fact-based explanations. Again, it is in those widely prevailing circumstances that upgrading to oil mist would prove highly valuable and easy to cost-justify [27, 28].

2.15.3 Oil mist provides more than just lubrication

Consider this portion of Chapter 2 an overview only. We want it to serve as a reminder that mist *lubrication* should always be mentioned together with the two terms *standby protection* and *oil mist preservation.* Because oil mist inevitably preserves stand-by equipment, the resulting reliability increase deserves to be reflected in the cost justification, as should failure avoidance and the ensuing reduction in maintenance outlays and, as an added bonus, pump-related fires [31].

Needless to say, this type of lubrication, protection, and preservation is even more easily justified in geographic regions with high humidity, or in regions with blowing sand. Additional credits are derived from oil mist lubrication for equipment drivers. Indeed, every experienced user plant applies oil mist to electric motor bearings [32, 33]. Oil mist is routed through bearings in accordance with the guidelines set forth in the eighth and later editions of the API-610 Standard (Fig. 2.27).

API-610 clearly depicts the optimized through-flow method that has been in use at some of the world's most profitable facilities since the late 1960s. Numerous papers and articles have documented this fact. Engineers from user companies, among them some of the largest multinational refiners and petrochemical companies, have freely shared their highly favorable experiences. Reliability professionals at these facilities are in the business of keeping plants running. At the same time, they have been tasked with finding cost-effective ways of extending and optimizing equipment uptime. Optimizing uptime does not mean adding maintenance cost and, in fact, implies maintenance cost reductions. The final outcome and ultimate test of a best-in-class facility has been, and will continue to be, lowest possible life cycle cost of all assets. In a great many places, oil mist lubrication has aided in meeting these test criteria.

2.15.4 Highlights and summary – 28 lubrication-related issues

A brief summary of upgrade items should start with bearing protector seals. Specify only the least risky rotating labyrinth-type bearing protector seal for pumps. Least-risk bearing protector seals will not incorporate a dynamic O-ring directly opposite a

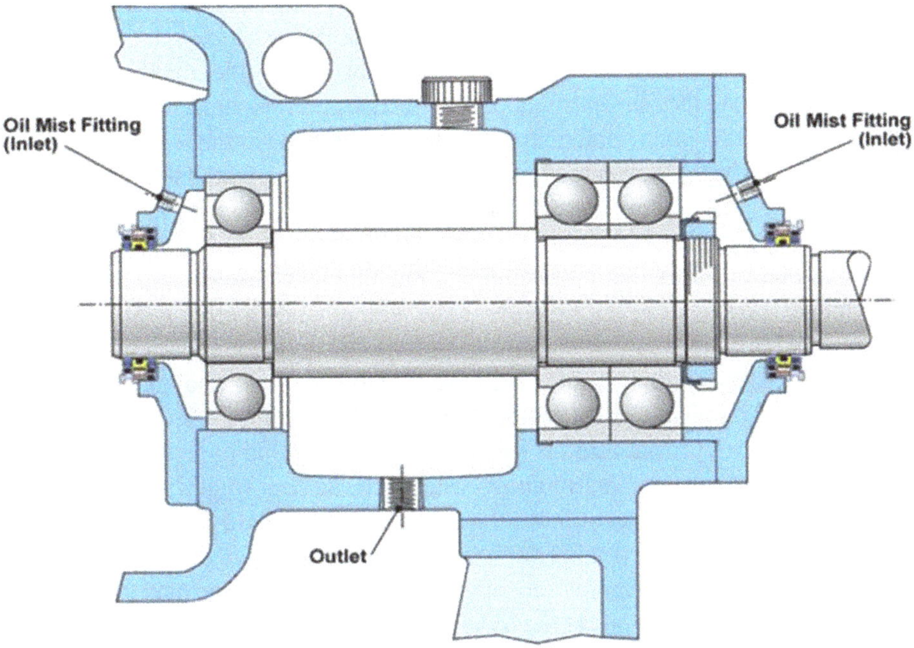

Fig. 2.27: Through-flow application of oil mist per API-610. Note magnetic seals.

sharp edge. The rotor of a low-risk bearing protector seal will be optimally secured to the shaft. Consider magnetic bearing housing face seals (Fig. 2.28) if the best possible oil retention and contaminant exclusion are needed.

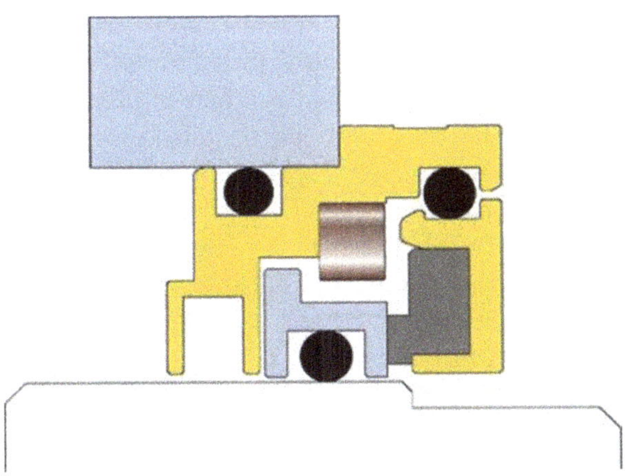

Fig. 2.28: Modern magnetic bearing housing seal for full, virtually hermetic, sealing of oil mist lubricated bearing systems. The suitably etched stationary face promotes face lubrication (source: AESSEAL Inc.).

Reliability professionals become thoroughly familiar with oil mist lubrication. Their employers represent the most profitable process facilities; many have used plant-wide oil mist since the early 1960s. Since about 1980 environmentally friendly closed oil mist units are available in sizes starting with two pump sets and ranging all the way up to 100 pump sets. A pump set includes the driver, usually an electric motor, ranging in size from 5 to 1,250 hp [34, 35]. The maintenance cost savings with oil mist are remarkable [36, 37].

With oil mist lubrication, vent openings, constant level lubricators, and oil rings are no longer needed; they are discarded. The maintenance requirements are sometimes reduced to a mere 10% of that needed for old-style traditional lubrication.

But checklists are still helpful in facilities that for some unexplained reason have not yet accepted oil mist. If constant-level lubricators are still used, ascertain they are mounted on the "up-arrow side" of shaft rotation. Do routine upgrading every time a conventionally lubricated pump enters your shop; be sure you retrofit a pressure-balanced constant level lubricator. Read the instructions carefully; some models with refill caps will malfunction if this knurled refill cap is not properly tightened.

1. On sump-lubricated pumps, do not permit oil levels higher than through the center of rolling elements at the 6 o'clock position.
2. More or thicker oil increases friction and heat generation [10].
3. On pumps with dn-values in excess of 160,000 (metric) or DN>6,000 (inches shaft diameter times rpm), allowing lube oil to reach even this position may result in excessive heat generation. For these, the oil level may have to be lowered further and oil rings or the superior flinger disks may have been chosen by the pump manufacturer and reliability-focused users [10, 37].
4. Except for moderate load applications in relatively cool ambient environments, ISO Grade 32 *mineral* oils and lighter lubricants are *not* suitable for rolling element bearings in centrifugal process pumps. A thicker ISO Grade 68 lubricant will perform better in the majority of rolling element bearings in pumps [23]. However, only a properly manufactured and flawlessly installed oil ring will perform satisfactorily in ISO Grade 68 and thicker oils. That is why ISO Grade 32 synthetics are often a logical low-risk compromise.
5. Bearing housing cooling may still be needed in pumps equipped with sleeve bearings. Most of these pumps require ISO Grade 32 lubricants; again, oil rings may not turn properly in the thicker oils. Close viscosity control may have to be maintained for satisfactory long-term lubrication [38].
6. Observe required ISO viscosity grades in moderate climates (Europe, the Americas, Pacific Rim, Australia): rolling element bearings – ISO 68, synthesized hydrocarbon optional; sleeve bearings – ISO 32, synthesized hydrocarbon optional; combining both bearing types in the same housing – ISO 32 PAO mixed with 10% dibasic ester synthesized hydrocarbon mandatory for extended life [39].
7. In bearing housings with both rolling element bearings (preferred lube viscosity is ISO 68) and sleeve bearings (which may require ISO Grade 32 lubricants),

consider satisfying both needs by using ISO Grade 32 or 46 PAO or diester-based synthesized hydrocarbon oils.

8. Next to oil-jet lubrication, "dry-sump" or pure oil mist applied in through-flow fashion per API-610/8th and later Editions represents the most effective and technically viable lubrication and bearing protection method used by reliability-focused industry. Use it for all rolling element bearings in pumps and electric motors [40].
9. Be aware of upgrade and conversion options whereby a simple and relatively inexpensive inductive pump (a small electric pump with a free piston the only moving part) can serve as the source of a continuous stream of pressurized lube oil. Used in conjunction with a spin-on filter, the resulting clean stream of lubricant can be directed at the bearing rolling elements for optimum effect [23, 41]. Placing a small oil pump in the bearing housing and driving it from the main shaft would be much preferred.
10. Certain grease formulations must not be mixed with other grease types. Incompatible greases often enter a chemical reaction that renders them unacceptable in less than 1 year [42]. If high-temperature PFPE-PTFE greases are involved, even trace quantities of any other grease will ruin the lubricating properties of the very expensive PFPE-PTFE product.
11. Over-greasing of electric motor bearings is responsible for more bearing failures than grease deprivation. Understanding and carrying out proper re-greasing procedures is essential for long bearing life [43, 44].
12. Lifetime lubricated (sealed) bearings will last for the duration of grease remaining in serviceable condition within the sealed cavity. Whenever the product of bearing bore (mm) times shaft rotational speed (rpm) exceeds 8,000, reliability-focused plants consider it no longer economical to use lifetime-lubricated bearings in continuously operating industrial machinery [23].
13. Grease replenishing intervals depend on bearing inner ring bore dimension and shaft rotational speed. Reliability-focused user plants consider dn = 300,000 the maximum for grease lubrication of electric motors and other machines in continuous service. It has been reasoned that beyond this dn-value (d = bearing bore, mm; n = shaft rotational speed, rpm), grease replenishing intervals become excessively frequent and oil lubrication would be considerably more economical [1, 21].
14. Realize that oil ring lubrication no longer represents state-of-art. Oil rings are alignment-sensitive and tend not to perform dependably if one or more of the following requirements are not observed [17]:
    - Unless the shaft system is absolutely horizontal (an unlikely event), oil rings tend to "run downhill" and make contact with stationary components. Ring movement will become erratic and ring edges will undergo abrasive wear.

- The product of shaft diameter (inches) and shaft speed (rpm) under well-aligned conditions should be kept below 8,000. Thus, a 3″ shaft operating at 3,600 rpm (dn = 10,800) would not meet the low-risk criteria!
- Operation in lubricants that are either too viscous or not viscous enough will not give optimized ring performance and may jeopardize bearing life.

15. Flinger disks (spools) fastened to pump shafts often perform considerably more reliably. Consider retrofits using flingers (or disks) made from solid stainless steel or with metal hubs/cores to which well-designed elastomeric disks are firmly fused or otherwise attached (see [8]).
16. If the use of oil rings is unavoidable, be aware that a 30° angle between the contact point at the top of a shaft and points of entry into the oil represents proper depth of immersion. Too much immersion depth will cause rings to slow down, whereas insufficient depth tends to deprive bearings of lubricant.
17. Oil rings with circumferentially machined grooves will provide increased oil flow but will still be subject to the concerns and constraints described in [17] and also in this text whenever oil rings are critically examined.
18. Ever since the 1960s, lip seals have been disallowed in API-style pumps. To achieve 6-year uninterrupted operation, either magnetic face seals or advanced rotating labyrinth seals should be used for bearing protection [11].
19. Consider near-fully sealed bearing housings to preclude ingress of atmospheric contaminants and egress of lubricating oil. Install two bulls-eye sight glasses on fully sealed pump bearing housings. Breather vents, oil rings, and constant level lubricators are no longer used in near-fully sealed bearing housings.
20. Note that a magnetic bearing housing seal approaches quasi-hermetic sealing.
    - Advanced rotating labyrinth seals will serve the majority of pump bearing housings, although all of these seals incorporate a small air gap. Note that at least one advanced configuration comprises two clamping O-rings when wider than 0.5 inches (~13 mm) and only one such ring when 0.5 inches or narrower.
    - Rotating labyrinth bearing protector seals configured with a sharp-edged groove directly opposite a dynamic O-ring risk degradation, oil contamination, and seizing ("lock-up"). This risk increases at slow-roll speeds and whenever there is relative axial movement between rotor and stator.
21. On grease-lubricated couplings, verify that only approved coupling greases are used. Do *not* mix different grease formulations [45].
22. Do not allow coupling greases to be used in electric motor bearings. Most of these bearings will fail prematurely unless a premium-grade "EM" grease (typically a shear-stable polyurea) is used [46]. High-temperature PFPE greases are often used in sealed bearings; however, they can only be cost-justified in bona fide high-temperature applications [14].

23. "All-purpose" greases are not suitable for electric motor bearings in reliability-focused plants. One cannot possibly become a "best-in-class" plant by "standardizing" on a single grease type [45].
24. Fully consider vulnerabilities of *unbalanced* constant level lubricators. Use only pressure-balanced models, such as TRICO, Oil-Rite, Lunkenheimer, and similar proven products [46].
25. If constant level lubricators are used, be sure to mount these on the correct side (the up-arrow side) of the bearing housing.
26. Whenever "up-arrow" shaft rotational direction is not observed and the constant level lubricator is mounted on the "down-arrow" side, there will be reliability issues. Mounting on the wrong side will lead to greater disturbances around the air/oil interface in the surge chamber of constant level lubricators. Correct mounting reduces the height difference between uppermost and lowermost oil levels. In other words, it ensures a more limited level variation [7].
27. On products that use caulking on the transparent reservoir bulb, recognize that caulking life is finite and that precautionary replacement (i.e., preventive maintenance) should be scheduled. Tiny cracks in aged caulking allow rainwater to be pulled into the oil via capillary action. Contaminated oil has a highly detrimental effect on bearing life [7].
28. Verify that re-lubrication and grease replenishment procedures [3] take into account that:
    - Mixing of incompatible greases will typically cause bearing failures within 1 year.
    - Attempted re-lubrication without removing grease drain plugs will cause the grease cavity to be excessively pressurized.
    - Considering elimination (discarding) of the grease drain plug and instead installing a permanent spent grease drain pipe [7]. This eliminates grease overload.
    - Over-greasing will cause excessive temperatures. On shielded bearings, cavity pressurization tends to push the shield into contact with rolling elements or the bearing cage, causing extreme heat, wear, and contamination.

Looking back on Chapter 2 and looking ahead to Chapter 3 we know that bearings and lubrication are subjects that cannot be neglected. Working with competent vendors, manufacturers and service providers is advantageous; the ultimate value of quality products and expert advice far outweigh their incremental cost. Reading good technical texts helps; we should consider it "icing on the cake."

With either pure oil mist or metered oil spray (both of which use an orifice-like spray nozzle or "reclassifier"), reduced amounts of frictional energy are created. The carrier air provides cooling. Because the small-bore piping leaves from the top of an oil mist header (Fig. 2.29), particulate matter, if it exists, will not be carried along in the oil mist. Instead, solids fall to the bottom of the header and only clean atomized oil arrives at the destination bearings. This observation has often prompted providers and users to label oil mist "the ultimate oil filter" [31].

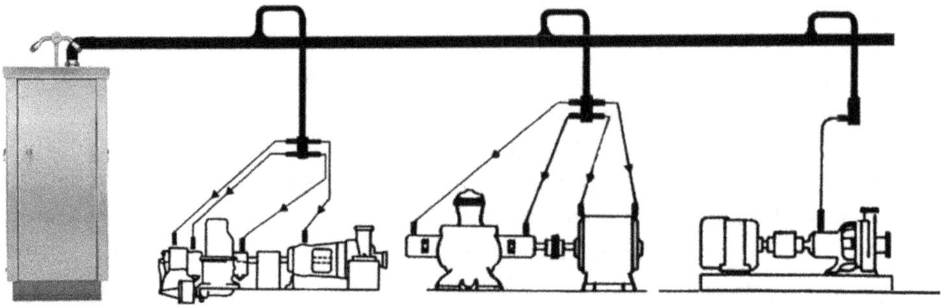

Fig. 2.29: Oil mist branch lines ("drops"] leaving from top of header; solid contaminants are left behind in the slightly sloped header and flow back to the console.

Ball and Roller Bearings [1] gives oil spray (also called jet-oil) ten out of ten possible ranking points and oil mist ranks a very close second [2]. The information and guidance in this chapter thus reverts back to the recommendations found in [4–6]; it harmonizes with the findings published by major bearing manufacturers. In other words, this chapter endeavored to answer the question: So what? The answer to this question is of significant importance and interest to many conscientious, reliability-focused industry professionals who recognize the shortcomings of lesser-ranked application methods.

We have learned that:

First: Avoiding failure risk is always the Best Available Practice

Second: The implementation of cost-justified upgrades is Best Practice, whereas "Business as Usual" and taking unnecessary risks is Inferior Practice

Third: Lubricant applications listed at the top of Tabs. 2.1 and 2.2 represent Best Practice.

At all times, working toward the acceptance and implementation of the top-ranked lubricant applications constitutes Best Practice for reliability professionals. It is only on rare occasions that one of the two top-ranked large-scale applications is not recommended for large plants with hundreds of process pumps and motor drivers.

But when would oil be the right choice over grease, and when should it be the other way around? We will attempt to answer this question in Chapter 3.

Chapter 3 General applicability ranges for oils and greases

3.1 Management digest

The primary purpose of lubrication is to separate stationary from moving parts by placing low-friction lubricant molecules between the two components. A flow of lubricant also serves to carry away heat. Oil has advantages over grease because it removes much more heat than grease. Grease has an advantage over oil because it is more easily confined to the bearing. But rarely is grease applied in accordance with best practices. Best grease application practices will be shown in this chapter.

3.2 Oil lubrication categories

Both oil and grease can be applied in different ways. Grease is normally used in electric motor drivers ranging from fractional horsepower to approximately 500 kW. This is because grease can be readily introduced in small-to-medium electric motor sizes where motor end caps are configured to accommodate grease. The end cap acts as a reservoir. A rolling element bearing retains grease and becomes a small reservoir if the grease is contained within shields that are pressed into the bearing's outer ring.

However, as bearing size and speed reach higher values, oil often represents an overall cost advantage when taking into account maintenance expense in calculating the total cost of ownership. A widely known, often used, rule of thumb links grease lubrication to machine size.

Oil applied as a static sump is often called an oil bath. Static sumps are acceptable for relatively low bearing velocities. With a static sump, the oil level is at or near the center of whichever rolling element passes through the 6 o'clock or bottom position. Oil bath lube is feasible for low-to-moderate shaft velocities. Once bearing elements plow through an oil bath at high velocities, heat generation becomes a concern. Elevated bearing temperatures can degrade lubricant oxidation stability.

3.2.1 Synthetic lubricants

Synthetic oil oils are synthesized hydrocarbons that (as a general rule) excel because of their high oxidation stability and low pour points. In other words, they allow higher operating temperatures than the considerably less expensive mineral oils.

For the most part, synthetics are made from polyalphaolefin base stocks. Dibasic ester stocks and various proprietary additive formulations are added in proportions that are rarely disclosed by the lubricant manufacturer. Additives allow tailoring and

https://doi.org/10.1515/9783110749441-003

fine-tuning the desirable properties of synthesized hydrocarbons. In general-purpose machines, the primary viscosities range from viscosity grade (VG) 32 to VG 100, the units being centistokes (cSt). VG 32 and 46 are often used in process pump bearings, although VG 68 predominates in very hot climates. VG 100 is trending toward gear oils.

Three important factors are worth mentioning here:

1. The overall coating and protection-imparting capabilities of a synthetic VG 32 are quite similar to those of a mineral VG 68,
   A four-fold extension of oil change intervals is often possible with synthetics.
2. As a rule, well-formulated synthetics will result in:
   - Corrosion protection for multi-metallurgy components
   - Lowering friction/stiction compared to competing mineral oils; this will reduce energy consumption by approximately two percent
   - Improved moisture protection
   - Better water separation
   - Minimized slippage of rolling elements
   - Reduced likelihood of sludge formation
3. Minimized inventory requirements
4. Most oil rings are designed and dimensioned to work with VG 32 and VG 46 lubricants. Operation with other VGs may be feasible but could invite problems.

3.2.2 Where synthetic lubes become important problem solvers

Synthesized hydrocarbons are long-chained molecules. The lube manufacturer/formulator must choose from a half-dozen base stocks and a seemingly inexhaustible combination of additives. Six machinery applications are listed where the higher cost of synthetics is justified over mineral oils:

1. **Automotive lubes.** Properly formulated synthetics are selected for their low friction and extended drain intervals. The greatly improved high-temperature tolerance of these oils very often mandates their use in high-performance engines.
2. **Rolling element bearings.** Over time, rolling element bearings tend to develop micro-cracks; vibration severity often increases as rolling elements contact the micro-cracks. Switching to an appropriate synthetic will fill the micro-cracks and vibration severity is thereby reduced or altogether eliminated.
3. **Gearboxes** and cooling tower fan gears will run cooler and live longer with good synthetics
4. **Closed-loop oil mist** lubrication units allow collection and re-use of oil over extended time periods. Significant operating and maintenance cost savings have been realized with synthetic lubes in closed oil mist units.
5. **Reciprocating compressor cylinders.** Here, too, greater uninterrupted run lengths have been documented with synthetics. Good formulations are often responsible for greatly extended valve and piston ring/rider band life.

Oil-flooded screw compressors benefit significantly from the many desirable lubricant performance uptrends achieved with synthetic formulations. Switching to superior synthetics allows higher compression ratios and, typically, fourfold run extensions before oil changes are called for.

Best Practices Highlights

With few exceptions, a fluid machine power output rating in excess of 500 kW favors two selections:

(a) oil lube application over periodic re-greasing and

(b) sleeve bearings over rolling element bearings.

However, just as there are many different oil application details, there also are different grease application details. These depend on bearing configurations; each merits further elaboration and will be discussed later in this text.

Suffice it to say that Reliability Professionals study the pros and cons and determine appropriate practices based on the outcome of these studies.

Chapter 4
Grease lubrication

4.1 Management digest

Grease lubrication is thought to have originated around 2000 BCE when the wheel hubs and bores of carts and wagons used by the Egyptians, Babylonians, and Assyrians received animal grease or tallow to remove frictional heat and reduce wear. Today's bearings use mineral oil-based greases, synthetic greases, aluminum-based, and certain perfluoropolyether–polytetrafluoroethylene (PFPE–PTFE)-based materials. The grease must be retained within either the bearing or its surrounding housing. Shields are used and shield orientation can be important.

One often differentiates between large and small bearings by looking at their respective dN-values. In this expression, "d" is the bearing's mean diameter in millimeters, N is its shaft rpm (revolutions per minute). Recall that in the-inch-and-feet system of measurements, DN stood for the bearing *bore* diameter (inches) times shaft rpm.

High values of DN or dN require more frequent relubrication than low values of DN or dN. Today's typical grease lubricated electric motor bearings are partially filled (never over-filled) with polyurea or other synthetic shear-stabilized "EM" (electric motor) rated greases. Grease classifications are usually expressed as NLGI (National Lubricating Grease Institute) grades. These range from a very fluid NLGI Grade 000 through a rather typical NLGI Grade 2 (multi-purpose) to a very hard NLGI Grade 5. However, favorable experience is reported with perfluoropolyether (PFPE)-based greases in fully sealed electric motor bearings. The PFPE experience completely upends prior notions regarding grease-filled or -sealed bearings.

There are only a few isolated instances when bearings should be fully packed with grease. A boat trailer is such an isolated instance. While the boat owner backs the trailer onto the boat ramp and submerges the wheels while launching the small boat, its owner wants to keep water away from the trailer's wheel bearings. When towing the boat on a highway, the bearings rotate at usually no more than 900 rpm. Let's assume that, on average, the owner tows the trailer two hundred hours per year.

Compare this to the average electric motor bearing. Its shaft diameter is twice that of the boat trailer's axle. The electric motor bearing rotates at twice or four times the speed and, hopefully, lasts 24,000 h, equaling three or more years. That's why grease in electric motor bearings should take up only 30–40% of the space between bearing rolling elements. Packing the bearing full of grease would create excess heat and significantly reduce bearing life.

https://doi.org/10.1515/9783110749441-004

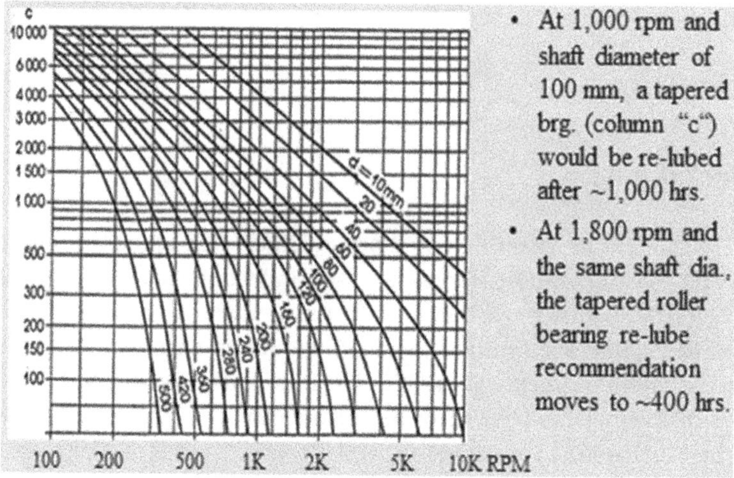

Fig. 4.1: Bearing relubrication chart intended for a particular bearing style (e.g., tapered roller bearing) as a function of shaft speed and shaft size (source: SKF America).

4.1.1 Grease relubrication intervals

Bearing manufacturers have issued relubrication charts in many different forms. On the example chart shown in Fig. 4.1 a particular bearing style is indicated by the letter "C" at the top of the vertical grid axis [45]. Depending on shaft size and speed, the recommended intervals can be read off on the vertical axis as hours between relubrication. While these intervals were conservative and pertained to standard greases, this type of relubrication chart was often used to envision where the use of lifetime lubrication – meaning fully sealed, non-regreasable bearings – should be discouraged. If a bearing cannot be regreased (and sealed bearings cannot be regreased unless we convert them to "open" bearings by discarding the seals), the indicated interval will at least provide a general guide on expected bearing life. The advent of entirely different non-hydrocarbon greases means that Fig. 4.1 does not in any way reflect the PFPE–PTFE grease Weibull plot line in the comparison plot of Fig. 4.2.

PFPE greases present an interesting lubrication alternative that was studied and fully validated at paper mills and similar facilities in the 2010 timeframe. It was determined that developments in grease technology can greatly extend the life of sealed bearings.

In one closely studied case, traditional motor bearings were supplied with sealed-in PFPE grease of the proper consistency (its PTFE ingredient is more commonly known as Teflon®). The resulting life extensions were tracked and explained on the comparison plot shown in Fig. 4.2. This detailed cost study at a Canadian paper mill showed benefits over periodic regreasing in certain industries and environments.

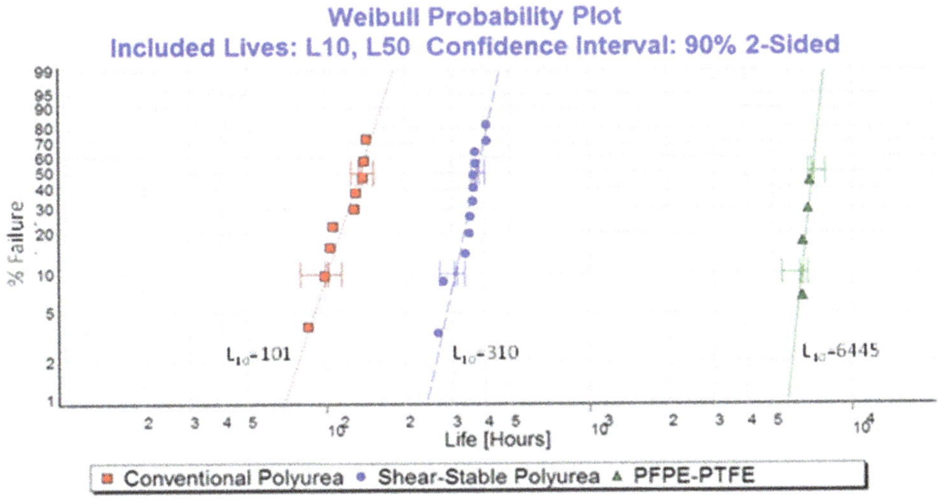

Fig. 4.2: Weibull probability plot for modern PFPE–PTFE grease formulations (source: DuPont, with data based on a Boulden Company newsletter).

However, the experience with PFPE–PTFE may not apply to every situation and careful follow-up is always recommended. Also, these greases cannot be mixed with even trace quantities of traditional grease types. Grease is therefore applied only at the bearing manufacturer's facility and only to absolutely clean bearings. Relubrication in the field has rarely, if ever, been successful.

4.1.2 Shields versus no shields in electric motor bearings

Through-path, unshielded, self-relieving bearings are shown in Fig. 4.3. Here, the motor manufacturer and designer ascertained that over-greasing is simply not possible. Excess grease flows into a spent grease cavity or to the ground. No shields are used in these self-venting bearings.

Chances are the through-flow design in Fig. 4.4 saves pennies over Fig. 4.3; however, spent grease will be expelled only if the drain plug is first removed from the bearing's housing. The human elements of training and conscientiously carrying out proper work procedures then become very important. Unless an escape path exists for spent grease, new grease being pumped in can cause extreme pressurization of the bearing housing. Grease temperatures can rise precipitously as the bearing rolling elements plow through a volume of grease that fills the entire available space.

Shielded bearings were soon adopted so as to reduce overfilling risks. A single shield is shown in Fig. 4.5; the shield is purposely located next to the grease reservoir to reduce the risk of over-filling the bearing. There is an annular gap of between 0.05

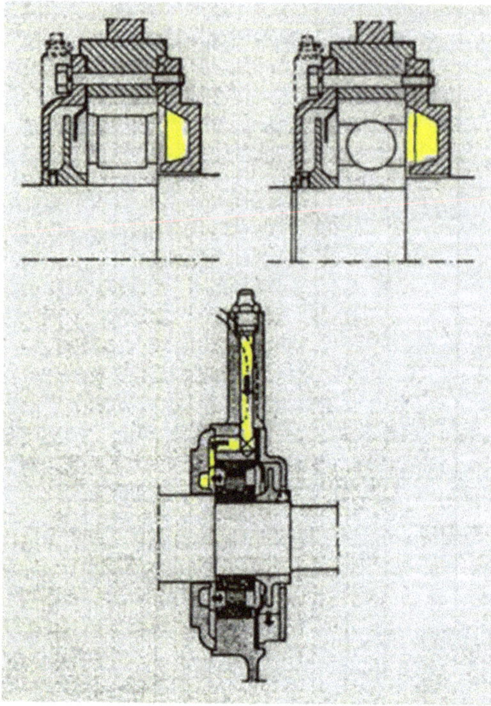

Fig. 4.3: Through-path, unshielded, self-relief bearing application range traditionally associated with lifetime lubrication in electric motor bearings.

and 0.10 mm (radial measurement) between the shaft surface and the shield's inner diameter. The design intent is for a small amount of oil to migrate from the grease reservoir into the bearing by capillary action. Grease replenishment would require removing the drain plug. Leaving it out for perhaps an hour before re-inserting the plug will allow the grease to flow in all directions. To re-emphasize, orienting the shield to face the reservoir will prevent packing (over-filling) the bearing with grease.

But design intent is one thing and field action is quite another. Inadequate training or personnel hoping to take shortcuts have often left drain plugs in place. In that case, excessive grease gun pressure often forces the shields into rubbing and scraping contact with rolling elements, upon which bearing degradation and failure become inevitable.

Double-shielded bearings (Fig. 4.6) at least eliminate questions as to which side should face the reservoir but leaving drain plugs in place still kills bearings. The industry has yet to devise a way that allows mechanics to put a new charge of unpressurized grease into the reservoir every few years and then let capillary action slowly move exceedingly small quantities of oil into the ball path. One manufacturer added a metering plate (red part in Fig. 4.7) in a valiant effort to ward off the over greasing calamity. The results were mixed; mechanics often took one look at the metering plates and threw them away.

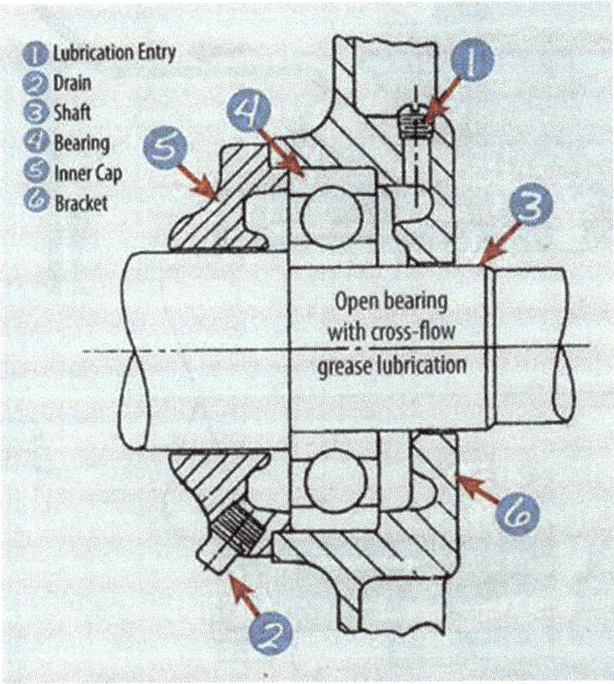

Fig. 4.4: Through-flow grease, where the drain plug must be removed while regreasing is underway.

ARCO Alaska, one of the legacy oil and gas producing companies originally partnering in the exploration of what soon became known as the Alaska North Slope, mitigated grease over-pressuring with a drain pipe shown in Fig. 4.8. ARCO's engineers, who added the drain pipe, left it open on purpose. As illustrated here, a small volume of spent grease forms a natural plug. The plug advances each time new grease is added to the reservoir. Sparing a description of what a heap of brownish grease looks like when it is expelled onto the pump base, it suffices to say after a year or so, it would have washed into a facility's oily water pit and can become more than just an eyesore for indifferent individuals and their employers. Nevertheless, let the record show that the ARCO Alaska method has proven to be both a reliability-enhancing and cost-saving modification that is routinely implemented by best-in-class companies.

Notice the similarity of Fig. 4.8 with a depiction in the earlier Fig. 4.3. Both leave the grease exit port open, regardless of whether the motor is mothballed or in operation. Still, Fig. 4.6 shows the best solution from a technically acceptable perspective. Having shields on both sides means a symmetrically manufactured radial bearing can be installed any way the machinist or assembler wishes. That is why two shields are also shown in Fig. 4.7.

Alternatively, one might institutionalize accountability and insist on staffers following instructions. One could teach the use of proper procedures and provide check-

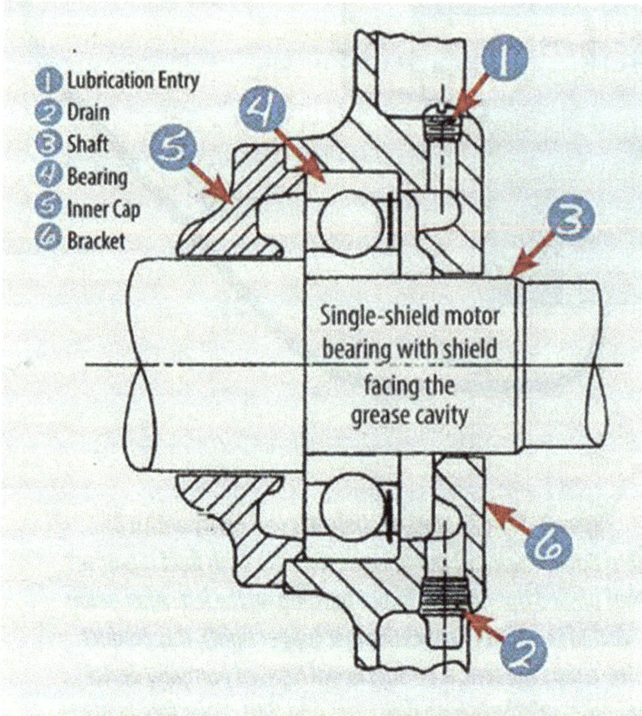

Fig. 4.5: Single-shielded bearing with removable plug.

lists. Insistence on the use of procedures is customary in aviation and is, in fact, mandatory. However, this mandatory approach works equally well in land-based facilities where safety and reliability are valued.

We know that the mandatory approach worked flawlessly in the UAE (United Arab Emirates) where a large refinery reported replacing seven bearings per 1,000 electric motors per year. When asked what magic grease formulation they were using, a senior manager explained that his workers simply followed instructions and that grease related bearing failures are a rarity. In other words, they knew what bearings they had, they removed drain plugs, they regreased with the prescribed amount of grease, and then they moved on to do the next electric motor. After allowing two to three hours for grease to settle, a worker returned and reinserted each drain plug.

The UAE experience simply illustrates that there is no substitute for following a proper work execution procedure. Good supervision and attentive management prevent failures and generate higher profits. How could this possibly be a novel finding?

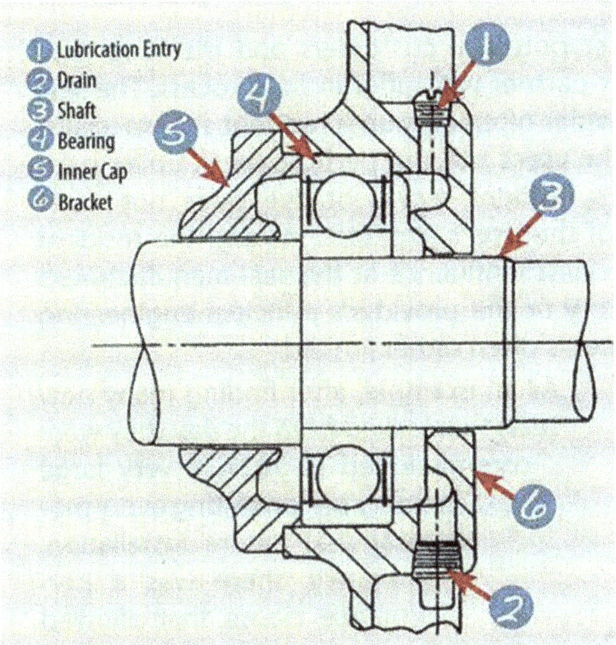

Fig. 4.6: Double-shielded bearings are sometimes used because their symmetry allows installation in either direction.

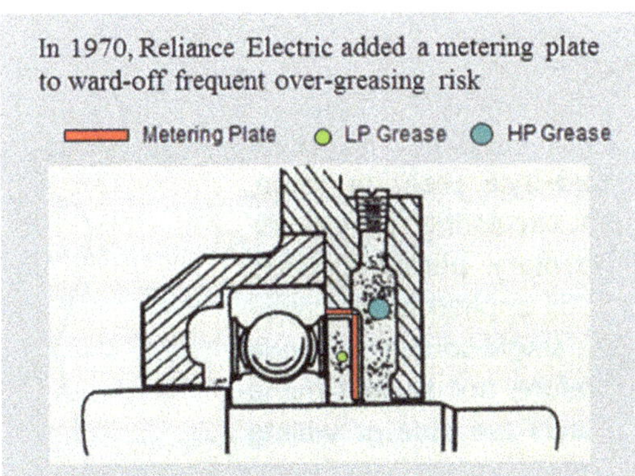

Fig. 4.7: Metering plate as the first line of defense against over-pressuring the grease (source: Reliance Electric, 1970).

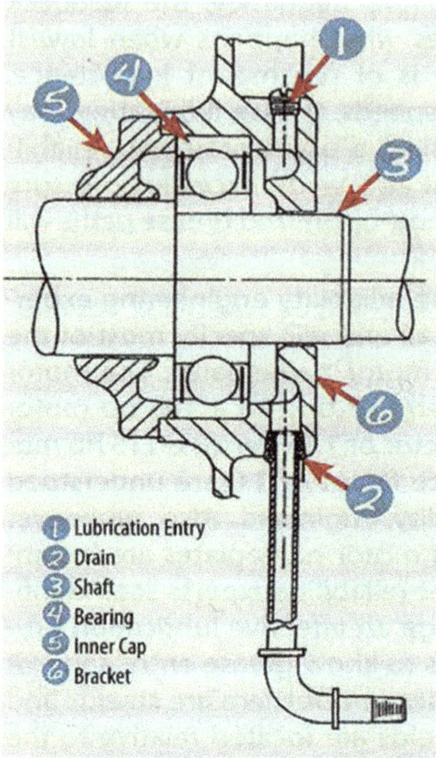

Fig. 4.8: Permanent drain opening prevents over-pressuring.

So then, here is what we have learned:

Best Practices would require purchase specifications to state more than motor frame size, voltage, insulation, and so on. Best-in-class companies specify bearing housing and grease flow details. Doing so avoids questions on regreasing with equipment running versus standstill, and questions relating to the merits of drain plug removal versus the risks of leaving drain plugs in place.

But instead of debating the point, we will look even more closely at "handicapped pumping equipment," in Chapter 5.

Chapter 5
Examining reliability-compromised process pumps

5.1 Management digest

Reliability suffers whenever best available technology is not being used. Motivated engineers are well aware of this; they will learn, disseminate, and teach facts. Facts enable these professionals to add value during their entire lives. However, as they continually update their knowledge base, they, no doubt, will experience a measure of opposition from peers whose energy output is misdirected, to put it mildly. Pump lubrication deficiencies are often denied, and the outstanding reliability performance of oil mist is vastly underreported and often mis-stated.

Beginning with 1950 and learning from aerospace activities, reliability engineers focused on industrial lubrication technology and studied the merits of synthetic lubricants. Without these synthesized hydrocarbons, aerospace endeavors would have moved at a snail's pace. Around 1960–1970, motivated reliability professionals began to familiarize themselves with the many different lubricant application methods found in process pumps. These professionals quickly learned that all manufacturers of rolling element bearings give ranking numbers to different lubricant applications. Then, like today, bearing manufacturers were unanimous in calling a jet of oil sprayed into bearings the most effective way of lubricating rolling element bearings. Oil mist (atomized oil droplets suspended and conveyed in very clean carrier air) is ranked a close second. However, there is a wide gap in the rankings between oil mist and the much lower ranked oil rings.

5.1.1 Revisiting "dn"

Please read on in spite of the overlap and repetition of a number of points we made earlier. Many equipment manufacturers have expressed shock when told that they provided an oil ring for dn (rolling element center-to-center opposite each other in mm times shaft rpm) in process pumps whose dn is as high as 1,100,000, although research by Wilcock and Booser [17] led to the stipulation that dn = 500,000 is the allowable limit. Nevertheless, oil rings are still widely used because they are inexpensive. To the wise, mere popularity rankings are as bad an indicator of reliability as lollipops are testimonials to good nutrition.

We wrote earlier that oil rings are a handicap for process pumps. For oil rings to function properly, they must be the right weight and have the right diameter. The shaft must be perfectly horizontal; the oil ring must be concentric within 0.002″ (0.05 mm); it must remain immersed in the oil a certain predetermined amount, and the oil must remain within a certain predetermined viscosity range. Reasona-

https://doi.org/10.1515/9783110749441-005

ble variations are allowed, although excessive deviations from one or more seemingly borderline but apparently “done this before” tolerances can be quite risky. If they happen at the same time, they will become the proverbial straw that broke the camel’s back.

Pressurized lubricant (e.g., oil jet or oil spray) is needed for the best possible lubrication of process pumps, but it would be offered only by the brightest manufacturers of process pumps. With a few notable exceptions, the use of external oil pumps (see Fig. 2.18) to pressurize and then spray lube oil into bearings is clearly uneconomical. However, the development and provision of a small oil pump internal to the bearing housing of a process pump is entirely feasible. Taking its cues from modern automobiles, a small pump could be located in or on the housing; it would be driven from the main pump shaft. The small pump would take flooded suction from the oil in the sump and force the oil through spray nozzles into the bearings.

5.1.2 Vendor response

Years ago, we tried to get vendors to understand that a confluence of all these requirements is rare. A highly skilled workforce would be needed and carrying out the proper installation and maintenance steps would require training, supervision, and compensation rarely found in the twenty-first century.

After a suitable presentation was made to a process pump manufacturer, its engineering staff studied the matter and advised they could easily design and accommodate a bearing housing-internal oil feed pump. The process pump manufacturer realized that such a pump would closely duplicate the oil pumps used in hundreds of millions of automobiles (typical as shown in Fig. 5.1). Such pumps are commonly driven by skewed gears at a few hundred rpm and, as the owners of modern cars well know, rarely fail. However, the process pump manufacturer realized that these small oil pumps would represent prior art, which makes them ineligible for patent protection. Fearful of others copying the design, the pump manufacturer decided not to market such products.

This does not mean, however, that a reliability-focused user must continue to put up with oil rings that have simply been tweaked from their previously flat cross-section to some newly advocated, trapezoidal cross-section, perhaps one similar to our earlier Fig. 2.25. Although oil rings made by combining high performance plastics (HPPs) with carbon fiber composites are considered more abrasion resistant than oil rings made of copper-containing metals, they are not the panacea. As mentioned several times throughout this book, it is time to implement reliable solutions rather than tweak 19th Century technologies.

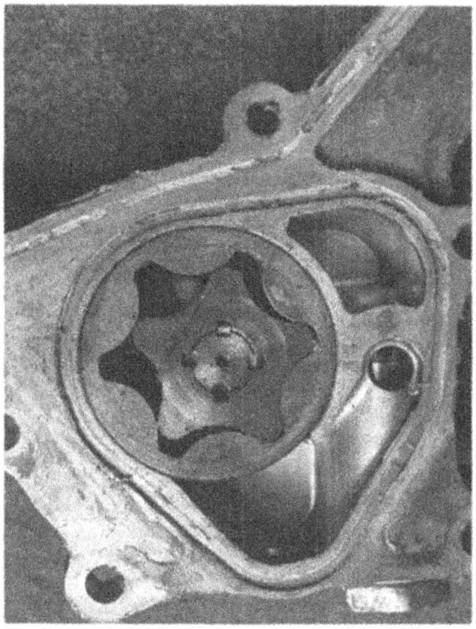

Fig. 5.1: Open source illustration of an automotive oil pump reliably driven by skewed gears (not shown).

HPP oil rings merely address one or more of the symptoms of the oil ring issue. They are not removing the underlying root causes of bearing life reductions that can have their origins in flawed oil rings. Among such underlying root causes we might find oil rings running downhill and contacting bearing housing interiors, oil rings that are not concentric to begin with, oil rings that have not been stress relieved by annealing (heat stabilization) and lose concentricity in operation, oil rings that are too light, too heavy, too deeply immersed in oil, immersed not deep enough in oil, immersed in oils of an incorrect viscosity or an untold combination of possible deviations.

Experience-based rankings of different lube applications were brought to the reader's attention earlier in this text (Tab. 2.2), and oil rings fared average, at best. Regrettably, nobody has data on how many oil rings will malfunction because the maintenance person did not properly install the shaft system, or the pump design allows the oil ring to slosh around; or the pump manufacturer supplied oil rings that were flawed for one or more of the causes listed above [17, 18]. The last thing a pump manufacturer would want its users to know is how many black oil incidents are caused by designs that overlook the need to provide an escape path for small amounts of oil trapped behind bearings (see Fig. 2.11) or for not seeing to it that there is pressure equalization between the space to the left and right of any particular rolling element bearing (Fig. 5.2). For grease-lubricated bearings refer back to Chapter 4 and keep in mind that every grease-lubricated rolling element electric motor bearing would run cooler, longer, and safer on pure oil mist. An oil mist-lubricated electric motor was commissioned in 1978 and was reportedly still running in a Texas plant in 2008. The owners claimed that no maintenance or changing of parts had been done in 30 years and an estimated 130,000 operating hours. Many moderately loaded ball bearings have L-10 catalog ratings (90% of the bearings not showing defect indications) of 500,000 and 600,000 h. Therefore, 30 years with standstill protection and vibration-free operation on pure oil mist simply meets expectations. In the case of eliminating grease lubrication, the vulnerabilities explained in Chapter 4 no longer exist with pure oil mist.

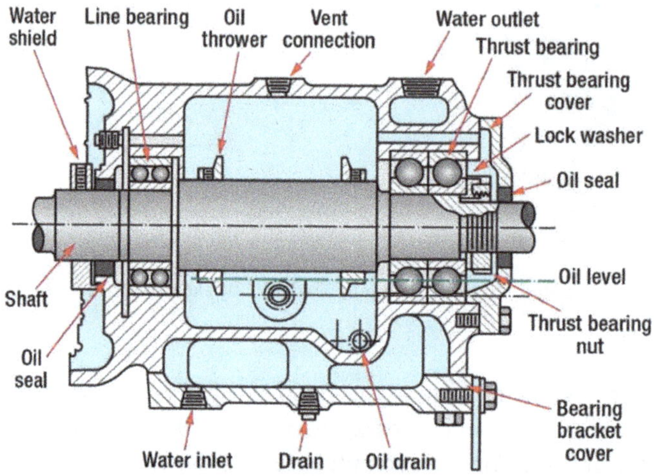

Fig. 5.2: Balance holes allow equalization of pressures on both sides of each bearing and the central part of the housing (source: Worthington Pump, Harrison; J circa 1960).

5.1.3 Prior art considered

The successful drive arrangement found in thousands of shaft-mounted team turbine mechanical governors (Fig. 5.3, left side) demonstrates relatively trouble-free shaft-driven pump arrangements. This would be but one of numerous spray pump drive options that derive from experience with automotive oil pumps and shaft-driven mechanical governors for small steam turbines. Except for a rather expensive, traditional, auxiliary lube pump-around system with a reservoir, an integrated lube spray system (see schematic representation with pump and filter, right side of Fig. 5.3) is the best of all worlds [23].

One no longer needs to worry about oil rings and their many demonstrated flaws, constant level lubricators, installation accuracy, shaft inclination, and so forth. Sooner or later, an innovative pump manufacturer will offer a pumping device and filter (see "P" and "F," respectively in Fig. 5.3) that will satisfy many thousands of

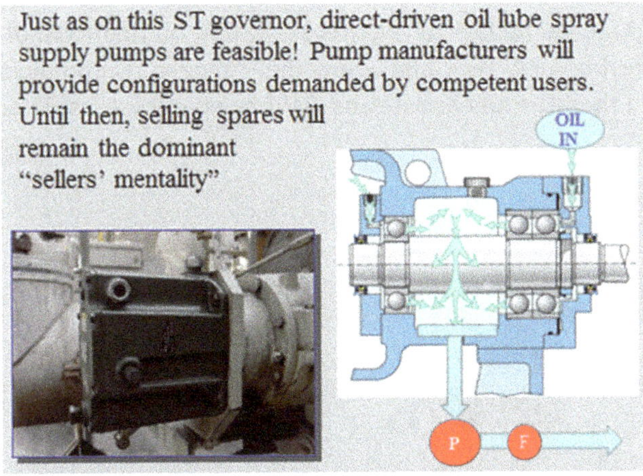

Fig. 5.3: Like the traditional, auxiliary pump around lube systems associated with mechanical governors (left), lube spray systems (right) are ranked best among all known lube application methods.

buyers. Process pump manufacturers unwilling or unable to supply oil spray might console themselves with purchasers who buy only from the lowest initial bidder or sellers of spare parts and maintenance services. It's a free choice everyone can make, yet responsible buyers favor sellers who add value, not uncertainty.

We refresh our memory by again considering Tab. 2.2 which was introduced earlier in this book (see page 40). Table 2.2 suggested a new pump lube application ranking; it was developed from data provided by a worldwide machinery engineers' network. During their twice-yearly meetings, these engineers examined data collected at their respective plants. They had access to at least 24,000 pump sets and decades of observing, studying, and cataloging elusive repeat failures. Elusive repeat failures still exist today, and this text is aimed at eradicating most, if not all of them.

No vendor is expected to agree with our occasional findings that their pumps fail more often than they should, and that upgrading risky components is in the buyer's interest. The vendor may be sincere in professing that he has never seen a such-and-such failure. However, decades of field observation by the user must be weighed against the manufacturers' test cycles lasting, typically, only a few hours. It appears these short duration tests became the basis for advocating plastic oil rings and somewhat thicker oils. However, these at best marginal improvements did not cure the problem of black oil experienced by a disappointed user company in Canada that made contact with the author within weeks of first reading about the issue in [16].

The findings and concerns published by Hooshang Heshmat and his co-investigator Oscar Pinkus [18] in 1984 cannot be disregarded. Even the highly relevant observations first documented in 1937 by researchers Baudry and Tichvinsky [12] give evidence that the behavior of oil rings is greatly influenced by a host of variables one encounters in field installations. A field-wise reliability expert will not be satisfied with efforts to tweak old components. Reliability experts know that oil ring technology is not the best available technology. They also know that tweaked components will require far more attention than the machinists and field mechanics are able or willing to give them.

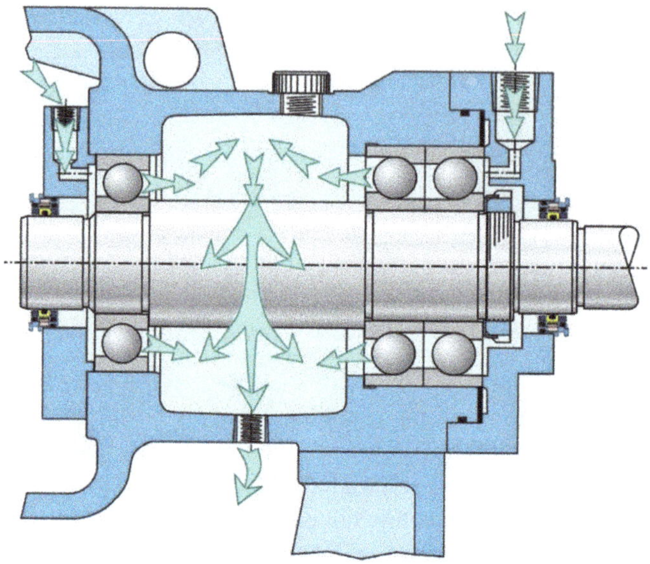

Fig. 5.4: Oil application via built-in "directed" nozzles.

We reached the conclusion that true reliability professionals would like to see pump manufacturers offer a built-in jet spray option similar to Fig. 5.4. The resulting better pumps would give rapid and substantial payback in return for their modest incremental cost increase over pumps provided with oil rings. The users and manufacturers would benefit from such pumps.

5.2 Why pump users should request lube delivery upgrades

Factory tests are valuable and should be welcomed. Both factory engineers and managers will agree that the machines on a pump manufacturer's test stand are properly aligned and the lubricant is fresh and clean. In contrast, the degree of inaccuracy encountered in many field environments differs greatly from the accuracy found on test stands. Neither the training nor the abilities of crafts and service personnel will measure up to expectations all the time.

In some installations, the piping connected to the pumps is pushing and pulling. As a result, housings deflect or yield, and bearings are edge loaded. The most elementary of equations explains what happens when a force of, say, 10 lbs acts on an original contact area of one hundredth of one square inch. Force per unit area equals psi, thus 10/0.01 = 1,000 psi, which is easily carried by an oil film. However, housing deflection may cause the bearing to contact the edge of the raceway, so, let's say only one thousandth of 1 square inch represents the area now contacted. If that were the case, the pressure would be 10/0.001 = 10,000 psi and metal-to-metal contact would probably result. The pressure would be far too high; the oil film would be interrupted and would no longer provide adequate separation of parts [10, 15].

Shaft alignment is often achieved by putting shims under the equipment's feet which, as a logical consequence, tends to cause shaft systems to be at a slight angle relative to the true horizon. On shipboard, pumps pitch and roll. Equipment surveillance and precautionary oil changes differ on ships from what is found at many land-based installations. These and other experience factors interact; they admittedly shape and skew rankings in the eyes of individuals with field experience. Their backgrounds and perceptions differ, so it is up to pump users to judge where to place their trust. We have observed that many process pumps in industry often experience repeat failures.

No reasonable person believes that one needs to show data to prove that driving cars with worn tires is a greater risk than driving on tires with tread. Also, what looks worn to one person, looks normal to another. It is no different with oil rings in pumps at location X versus location Y. Gravity being gravity, the sketch shown earlier in Fig. 2.25 is a scientific fact of the effects of gravity and friction. The ultimate ramifications of a trapezoidal oil ring operating in this manner can be foreseen: A trapezoidal oil ring has two pointed or circumferential ridges, one on the left side and one on the right. Pointed ridges have a very small total surface area. As the oil ring slews from side to side in its carrier sleeve, it touches the side of the carrier sleeve. This means the force per unit area, the pounds per square inch (commonly known as pressure) will be rather high. When that happens, the pointed ridge will break through the oil film and abrasion will occur.

Whenever a pointed ridge breaks through the oil film, there is increased friction, and the oil ring slows down. Long-term satisfactory operation is at risk unless the pump owners invest heavily in preventive maintenance action. But preventive maintenance costs money, and that is simply an additional reason why oil rings rank average to low on the pump lube application ranking table.

Again, jet oil lubrication [19] represents the highest rated application method. Also called oil-air, see Fig. 5.5, jet oil lubrication has been widely used in aerospace since 1949. It can open a window of opportunity for reliability-focused users and/or innovative pump manufacturers. Think of a small oil pump ("P" in Fig. 5.3), either internal to the process pump bearing housing or incorporated in a small assembly screwed into the bottom drain of a process pump. This upgrade can provide filtration, metered flow, and proper pressure downstream of the oil sump and upstream of the spray nozzles. Motivated users have written this preferred lube application approach into their pump specifications and are actively pursuing this pump reliability enhancement.

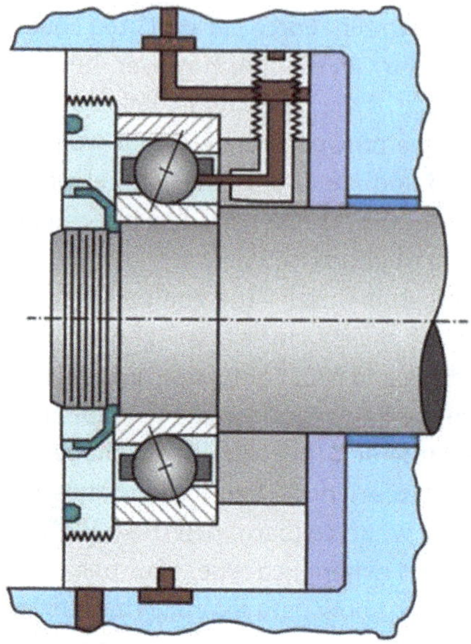

Fig. 5.5: Jet oil lubrication, whereby liquid oil is sprayed (and directed) into the rolling elements of a bearing [5, 19].

Having examined process pumps which come with reliability compromises, it is only fair to state that things do not have to be that way. More to the point: Whatever style oil rings users *presently have* in their pumps are probably not the most reliable components. More dependable means of applying lubricants to bearings are either available or have been designed. Intelligent upgrading would reduce maintenance labor and improve the safety and reliability of many plants. The recipients of better pumps would suffer fewer pump fires (see Fig. 5.6) and would operate more safely, consistently, and profitably.

Meanwhile, advocates for increased pump reliability attempted to cajole a few manufacturers into offering upgrades. The next subheadings highlight the status as of late 2018 and early 2019. As of 2021 we believe there has been no change in status and risky oil delivery methods are the low-cost normal.

5.2.1 Slow progress in obtaining lube delivery upgrades

As the first edition of this book went through its final editing process in mid-2019, the responses from several pump manufacturers ranged from no reply to claims that oil rings work just fine. One pump manufacturer had never heard of jet oil and asked us to be more consistent in our notation and to call it oil mist, as we had done in articles and books since 1972. Another pump manufacturer opined there was no user demand for built-in jet oil lube provisions.

Fig. 5.6: Relating pump repair events and fire incidents.

But the specifics of three more extensive contacts – we'll call them X, Y, and Z – are even more interesting. Here is what transpired when we took the time to supply X, Y, and Z with more formal requests.

5.2.2 The gear pump meeting

A full day visit to a reputable pump manufacturer included a presentation on the merits of jet oil lubrication. As the meeting came to a close, representatives from manufacturer X repeated their understanding and advised that bearing housing internal gear pumps (see Fig. 5.7) would be a relatively easy and cost-effective solution. From our own studies, we concluded that small gear pumps would be about 1.5″/38 mm thick and have perhaps a two-by-four inch (50 × 100 mm) or smaller footprint. Gear pumps are very efficient, quite inexpensive, and could be driven by skewed-tooth right angle bevel gears or parallel spur gears. Worst-case scenarios would call for modifications to existing pump bearing housings. A few existing pump models incorporated housing dimensions and configurations that seemed ideal for the suggested upgrades.

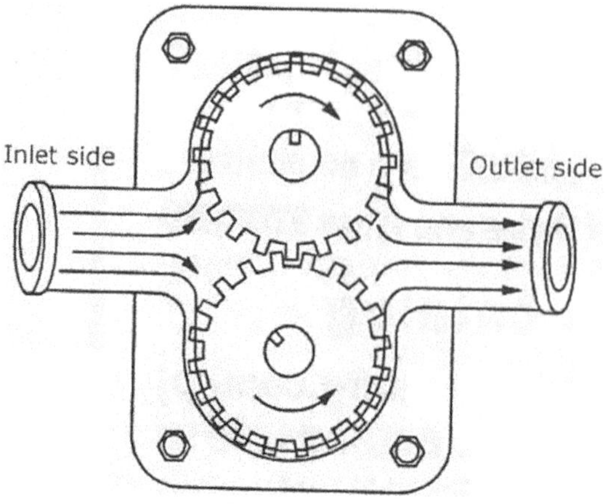

Fig. 5.7: A gear pump could be placed in the oil sump and one of the two gears fitted with a driving shaft connected to the main shaft.

However, representatives from the widely known manufacturer X later advised that their straightforward, cost-effective, and reliability-improving solution could not be patented and, thus, would draw imitators from overseas. Therefore, they regretted to inform us that they will have to decline any further involvement.

5.2.3 Pursuing regenerative pumps

The manufacturer of well-engineered small regenerative turbine pumps (manufacturer "Y") expressed near certainty that one of its smallest size pumps would be well suited for oil oil jets. "Jetting" is another name for directly aiming pressurized oil through a filter and small spray nozzles into the pump's bearings. However, we received neither an answer to our initial conversations nor was there a reply to our follow-up attempt. We were surprised that not even an e-mail reply followed an hour-long conversation with company Y's CEO. We had considered these conversations quite promising at the time.

5.2.4 The 26 pump lube improvement opportunity

A well-informed pump user in the United States was quite aware of an advertisement published decades earlier by a noted US manufacturer who had a solution for oil ring problems. This manufacturer offered a pump with a disc (they called it an "oil thrower") mounted at the shaft end and partially immersed in a generously sized sump (Fig. 5.8). The pump user also noted a breather vent to the left of the shaft-mounted oil thrower and a slanted vent hole or pressure equalizing hole on the right side of the thrust bearing set. Such vent provisions are important signs of thoughtful designs.

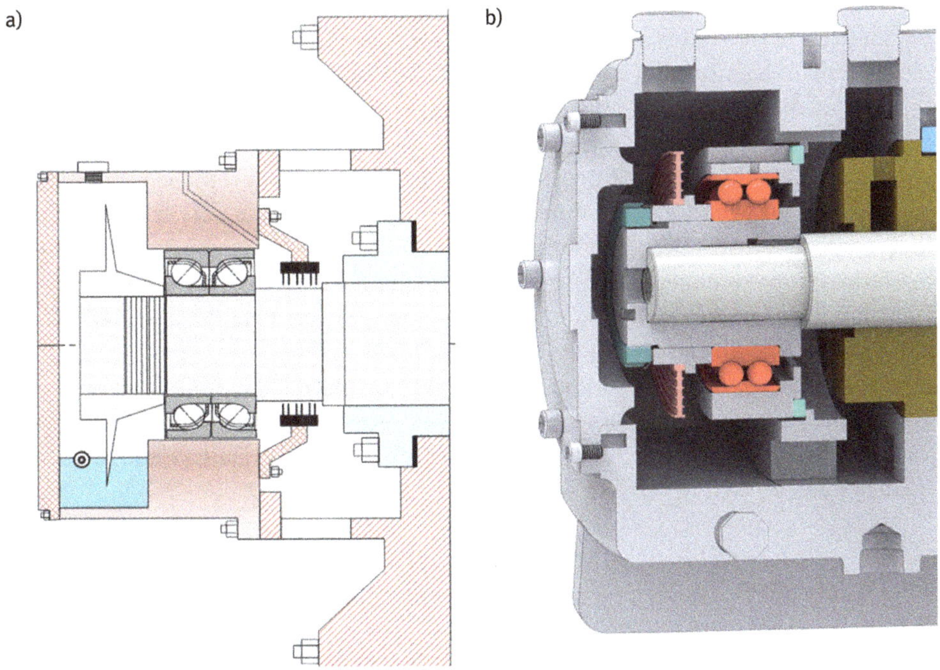

Fig. 5.8: (a) The 1960s advertisement clearly conceding that an "OIL THROWER ensures positive lubrication and eliminates the problems associated with oil rings." (source: Hurl Elliott). (b) One more "mockup" of the 1960s advertisement clearly conceding that an "OIL THROWER ensures positive lubrication and eliminates the problems associated with oil rings."

In the early 1970s, this legacy pump manufacturer had attempted to sell solutions to "eliminate the problems associated with oil rings" [7, 10]. This convinced the well-informed pump user to develop inquiry specifications in 2012 that included an oil jet lube application requirement. The specification, which was part of a project involving 26 pumps, was sent to 4 bidders. Every one of the four bidders took exception to the requirement and offered their standard products only. Pressed for time, the specifying owner-purchaser's project engineers did not want to search for pump manufacturers whose attitude would be more receptive to requests from the plant's reliability professionals for pumps with the improved lubricant application. An opportunity was missed, and one will never know how many requests from other reliability-focused potential clients might have followed.

By way of simple reaffirmation and reassurance, the various lube application methods presently found in pumps and electric motors will work. However, one may question the motives of a pump manufacturer who, by tweaking old oil rings, achieved little, if any, improvement. There has been close to zero innovation from some manufacturers, which should prompt us to question whether they really deserve our business.

Reliability professionals may advise project managers to steer clear of sending bid invitations to unresponsive pump manufacturers. Purchasing equipment from manufacturers who understand what it means for the purchaser to request more reliable equipment is essential if we pay more than mere lip service to the term "reliability."

One of the top corporations in the petrochemical and refining industry accomplished the goal of higher asset reliability; their growth and promotion policies achieved long-term results. The corporate heads informed project managers that their next promotion would be to plant manager of the facility or process unit which was their responsibility to design or construct right now. It was common knowledge that the future plant manager would be leading the plant for at least 3 years before being considered for another promotion. The project manager now had a personal interest and stake in building a safe, reliable, and profitable facility.

Meanwhile, pump users often cannot wait for the design and manufacture of better pumps. There are instances where users must cope with and address issues with great urgency. The next subheading demonstrates and relates these facts with relative ease.

5.2.5 Implementing an immediate upgrade in the Western United States

In April 2015, we received a distress call from the maintenance manager at a midsize refinery in a US western state. The refinery was in the process of starting up a new process unit and two of the new pumps were failing. We later concluded the failure was the result of a serious design error. The pump bearing housings were wrongly dimensioned and the oil rings in both pumps repeatedly failed within days. An immediate solution had to be found but pursuing any of the time-consuming upgrades was out of the question.

There were two options: First, the refinery could contact the three mechanical seal manufacturers whose product slate included small, stand-alone external pump around units (Fig. 2.18). Pump around units typically move and/or pressurize barrier fluid in dual mechanical seals. They could perform as jet oil application units, but their use would require installing spray nozzles within 12 mm (7/16th inch) of each of the two pump bearings.

A second option would be to locate, purchase, and priority ship the small, self-contained oil mist generator/oil reservoir package depicted earlier in Fig. 2.19. A pre-owned package was immediately located; it even included several oil mist reclassifiers (most of these are similar to a slender metering orifice with an L/D ratio greater than two) for installation at or near the bearings. We assumed the package was traded-in by a pump owner who had recently purchased and installed a larger capacity oil mist generator unit, perhaps for plant-wide use.

But the real point of the story deserves our attention. The pump manufacturer had most probably developed a new pump with a bearing housing of insufficient volume. Moreover, there were indications that the manufacturer had compounded his error by providing oil rings that were deficient in more than one detail.

The small oil mist generator arrived at the midsize petroleum refinery within 2 days. Hooking it up took an experienced oil mist installer a few hours and the problem was solved. Had the refinery initially insisted on pumps without oil rings or acted on the first indication of issues with the new oil ring equipped pumps, the owners would have avoided cost issues and losses in the seven-digit range.

Oil mist provided a ready solution to the problem at this mid-western oil refinery and similar successes exist at hundreds of other locations. Oil mist technology is well worth knowing about and Part B of this text explains the underlying principles. Part B also describes and quantifies the intrinsic value and cost advantages of this very mature lubrication technology. Well over 3,000 oil mist generator cabinet installations exist at hundreds of facilities. They serve the plant-wide and individual reliability needs for highly dependable lubrication. And oil mist systems of astounding simplicity have also proven to be the most effective method of storing equipment for virtually unlimited periods of time.

We have learned that:

Best Practices Companies have developed supplemental pump specifications and attach these to API (American Petroleum Institute) Standards, in particular API-610. The specification amendments or addenda stipulate upgrades and make it difficult for bidders to take blanket exception to the various upgrade requirements.

Best Practices Companies are willing to pay for upgrades but will explain to pump manufacturers the evenhandedness of distributing the cost of reasonable upgrade designs over an anticipated number of future sales. Manufacturers unwilling to see the wisdom and mutual benefits of these requirements should not be favored with orders from reliability-focused purchasers.

Best Practices Companies pursue growth and promotion policies that instill in project engineers and managers a personal interest and stake in building safe, reliable, and profitable facilities.

And that gets us to Part B of this book.

Part B: **Fundamentals of oil mist technology**

How equipment outdoor preservation later becomes full standby protection

Chapter 6
Oil mist technology and its role in optimally protecting standby (standstill) equipment

6.1 Management digest

First off, it is important to assert that Chapters 6 and 7 do not infringe on or limit the lube application choices made by manufacturers and/or equipment users. All kinds of lubrication are presently available. Some are labor-intensive and relatively uneconomical, but they do the job. Others require a highly trained workforce and, if such a highly trained workforce is carrying out the task, these existing application methods will also work. However, Chapter 6 in particular sets out to address the needs of reliability professionals and other groups that are fully cognizant of the flaws seen by Eschmann, Hasbargen, and Weigand and implied in the rankings reproduced earlier in Tab. 2.1.

Therefore, this chapter provides solid answers to professionals who are unhappy with a presumed 90% or, perhaps, even a 95% reliability achievement record. There is evidence that best-in-class (BiC) users are achieving 99.5% of possible run length and, accordingly, that should be the goal. Putting it another way: Facilities where critical machines experience 4.5% more failures than their competition don't stand a chance of joining the small group of BiC performers. If the Fire Engines dispatched by City "A" reach their intended destinations 995 out of 1,000 times and those in City "B" arrive at fires only 945 out of 1,000 times, the differences may be significant in terms of asset preservation and lives endangered.

While traditional methods can work, they simply do not represent best available technology. Superior ways to reduce pump-bearing failure risks and frequencies exist and are described in several of the later chapters in this book. In Part B, the emphasis shifts from the operating life extension of fluid machines and electric motor drivers to storage preservation and protection before equipment commissioning. But Part B also gives more detailed cost comparisons and other explanations dealing with oil mist lubrication, which is one more reason why consideration of Parts A, B, and C should be given equal weight.

6.2 Brief overview

Broadly speaking, optimized *lubrication delivery* for operating machines is quite similar to optimized *storage preservation*, also called "equipment mothballing." Oil application in mist form (as an "oil fog") and/or in liquid spray form is considered

https://doi.org/10.1515/9783110749441-006

optimum by the most renowned bearing manufacturers. If initially stored and later lubricated with synthetic lubricants, the commissioning time is minimal and oil compatibility is ensured. Because of their relative importance, reliability-focused owners and users consider lubricant types and the details of lube application as top items of interest.

Oil mist consists of atomized globules of lubricating oil suspended in clean, moisture-free, instrument-quality air [46]. The mixing ratio is 200,000 volumes of air per volume of oil. With its concentration, thus, only 0.05% oil by volume and somewhere near 0.1% by weight, it is considered orders of magnitude too lean to be a safety concern. It neither can explode at the prevailing temperatures, pressures, and concentrations nor can it support a flame. After the oil and air are combined in either a vortex-generating module (Fig. 6.1, right) or a more widely used ejector-type venturi module (Fig. 6.1, left), the module is mounted in a suitable enclosure, console, or cabinet (Fig. 6.2). Once air and oil are mixed, the resulting "oil fog" or "oil mist" will have the appearance of cigarette smoke or hair spray.

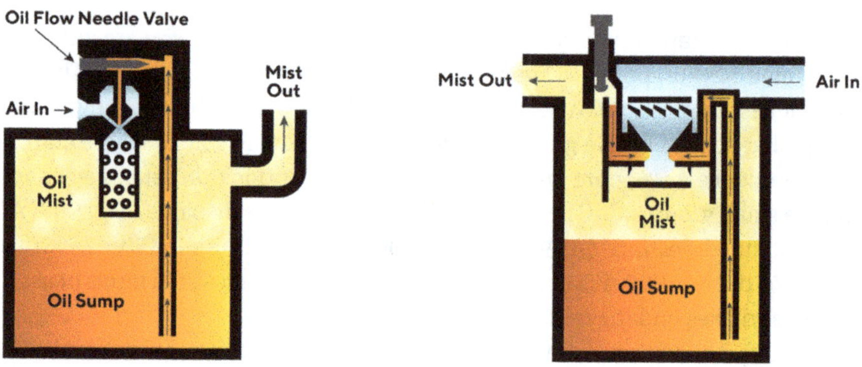

Fig. 6.1: Air meets oil in either a vortex generator (right) or a venturi-type nozzle (left) (source: T.F. Hudgins, Houston, TX).

On its way to the destination, oil mist is conveyed in headers (Fig. 6.3). Branch lines take off from the top of headers; they terminate in a manifold designed with typically eight threaded outlets. Individual stainless-steel tubing lines connect to these manifold outlets. Oil mist flows through a reclassifier nozzle into an operating bearing (Fig. 6.4). The flowing oil mist then encounters turbulence, and the small (atomized) globules combine into larger oil globules. These globules are then too large and too heavy to remain suspended in air; they combine into droplets that coat all bearing elements with oil.

One of the two major oil mist providers uses stainless steel for enclosures and consoles of various sizes (Figs. 6.5–6.7). The other provider offers a choice of stainless steel and aluminum.

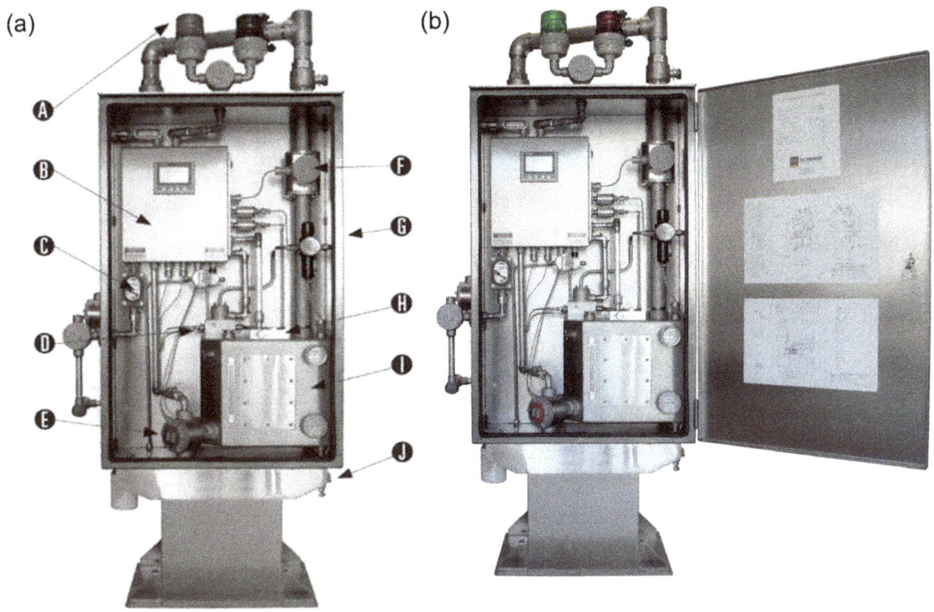

A. High visibility LED red and green operating status lights

B. Human Machine Interface (HMI) display of operating variables with red and green alarm indicators

C. Electrical power switch and terminal enclosure with common remote alarm contacts

D. Air heater w/high and low-temperature alarms

E. Oil heater and temperature gauge w/high and low-temperature Alarms

F. Oil Mist Density Monitor

G. Inlet air with integral air filter/regulator with gauge and auto filter drain

H. Mist generator head - 40, 100, 250, 500 or 1,000 BI sizes

I. 9-gallon stainless steel reservoir with pressure relief valve, oil level sight gauge and clean out panel

J. Stainless steel cabinet

Fig. 6.2: Oil mist generator console/cabinet (option 1) (source: T. F. Hudgins, Houston, TX).

Fig. 6.3: The two oil mist headers in this photo terminate in the upper right foreground; branch lines or "drops" go to a small manifold (source: Don Ehlert).

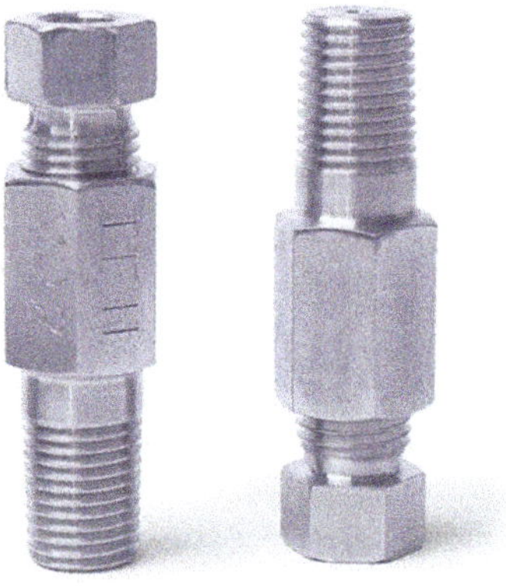

Fig. 6.4: Oil mist nozzles (reclassifiers) (source: T.F. Hudgins, Houston, TX).

6.2.1 Coalescing action

It is easy to visualize that the rotation of the bearing elements promotes collision of many thousands of atomized oil droplets each second. They collide with each other and with the balls, rollers, or cylinders that rotate in a bearing. Oil droplets and the

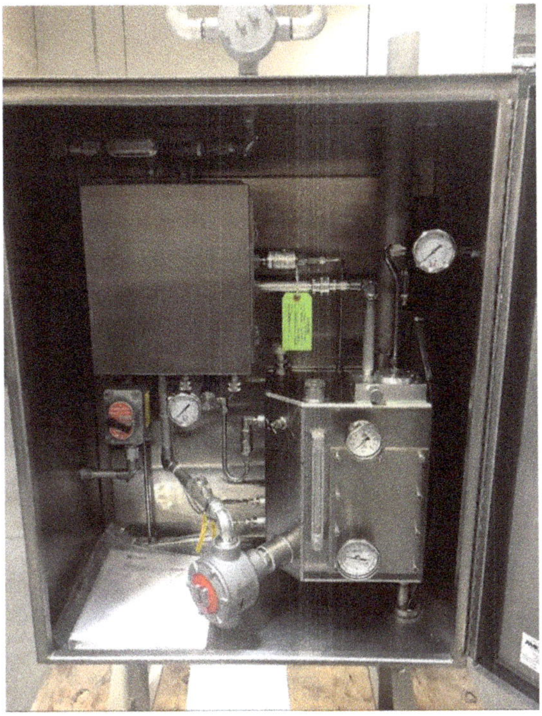

Fig. 6.5: Medium-sized oil mist console, open view (source: T. F. Hudgins, Houston, TX).

remaining atomized oil globules then plate-out and coat, or "wet out" on the bearing elements. This oil coating establishes itself as a separating layer between rolling and stationary bearing parts. With the mist, thus, having lost most of its oil, the remaining carrier air continues toward a convenient down-facing vent hole or exit port, all the while removing heat from bearings and bearing housings. Bearings lubricated by pure oil mist typically operate 15–18 °F (8–10 °C) cooler than bearings with oil applied by one of the many traditional methods.

Because *optimally* protecting the equipment involves oil mist technology, the next few sections and chapters endeavor to explain this technology in greater detail. Once readers have a grasp of oil mist technology, they will find it quite easy to see its linkage to the protection of machines on standby, and the preservation of stored (mothballed) equipment. In particular, parts of the present chapter will again serve as a form of "Management Digest." The chapter could also provide insight into the reasons why no leading BiC oil refinery has ever reached this coveted designation without extensive use of oil mist lubrication.

Fig. 6.6: Medium-sized oil mist console, external view (source: T. F. Hudgins, Houston, TX).

6.2.2 Lubrication volume and reclassifier sizes

The sizing data covering all rolling element bearings with bore diameters from 3″ 75 mm) to 20″ (500 mm) are listed in Tab. 6.1; however, other sizes are also used. The desired flow volumes (including generous margins of safety), bore diameters, and bore lengths of the five most often used mid-range reclassifiers are listed on this table. The smallest of the most frequently used bore diameters is 0.032″ and its effective bore length is 1/4″ (approximately 6 mm). The physical length of the reclassifier may be whatever convenient length the provider chooses, but most of that length will be opened up or over-bored, that is, only 6 mm of the length is the active bore diameter. The need for specially dimensioned reclassifiers is infrequent.

Assuming a header pressure of 20″ of water column (~5 kPa) 0.09 scfm (standard cubic feet per minute) would be the volume of oil mist flowing through this reclassifier. This would be the flow volume that satisfies the lubricating and cooling needs of one nominally 3″ (75 mm) shaft diameter bearing. The term bearing-inch ("BI") denotes the product of multiplying the number of bearings and their respective shaft diameters. Three BI could be 3 bearings with a 1″ diameter, or two bearings side by side with a 1.5″ diameter, and so forth.

Fig. 6.7: Small self-contained oil mist unit suitable for lubricating two or three pump sets or other equipment in a small storage yard. The blower mounted on the returned (coalesced) oil container produces a slight negative pressure (source: T. F. Hudgins, Houston, TX).

Tab. 6.1: Typical reclassifier sizes and

Typical reclassifier data			
SCFM rating	**Bearing inch rating (BI)**	**Bore diameter**	**Bore length**
0.09	3	0.032	1/4"
0.18	6	0.047	1/4"
0.30	10	0.060	3/8"
0.45	15	0.073	7/16"
0.60	20	0.086	1/2"

Let us use one more simplified example and assume we had two 9″ (230 mm) bearings mounted in tandem. Because 2 × 9 = 18 BI, we would select the 20 BI reclassifier and send 0.60 scfm of oil mist through a 0.086″ diameter bore into the bearings. And always remember that every conceivable type and style of rolling element bearing has been lubricated by pure oil mist. For very slow speeds, the reclassifier bores may have been "rifle-drilled" to promote coalescing at the bore exit, or several small reclassifiers may be used instead of just one large one.

6.3 Oil mist technology and its role in optimally protecting equipment

It has long been proven that soundly executed long-term storage protection of sensitive machinery, such as pumps, turbines, compressors, mining machinery and drivers, will save much money. The same is true, of course, for general-purpose ("GP") machines, the bulk of these being process pumps, electric motors, and all sizes and types of steam and gas turbines. Numerous other pieces of equipment including vessels, drums, columns, and so forth have benefitted from storage protection.

6.3.1 "Mothballing" and how it works

Among the many examples that illustrate the importance of storage preservation, a decommissioned and subsequently storage-protected (i.e., "mothballed") fertilizer plant in the southern United States serves as an interesting standout. The plant sat idle for a considerable time before it was purchased, dismantled, and shipped to Pakistan, where it was reassembled and recommissioned in the early 1990s. Many years later, the plant had become a model facility. Using natural gas as its primary feedstock, the facility was manufacturing fertilizer at a fraction of what it cost the competition to produce. This plant's total investment was far lower than the cost of a completely new plant. Good storage protection had greatly contributed to its remarkable financial success.

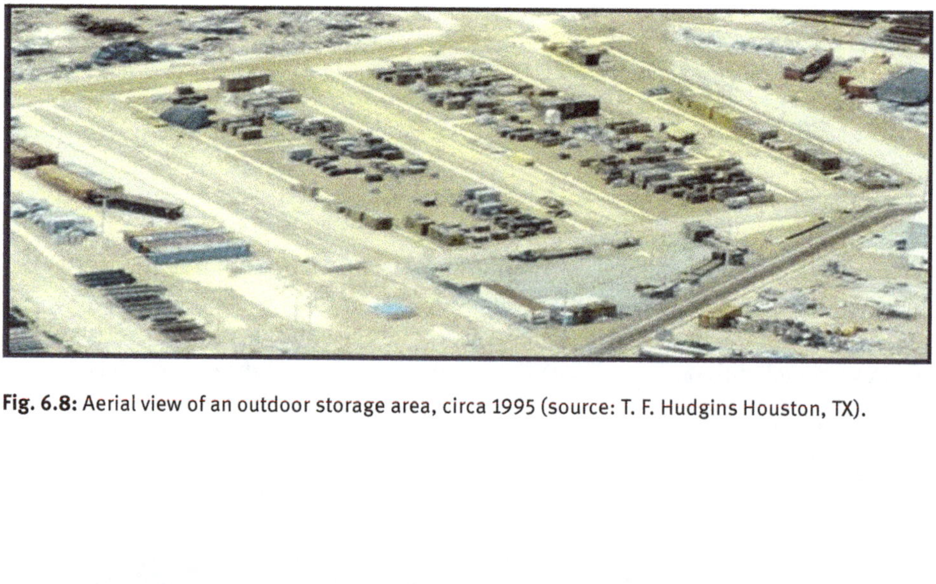

Fig. 6.8: Aerial view of an outdoor storage area, circa 1995 (source: T. F. Hudgins Houston, TX).

6.3.2 No downsides, only advantages

The secret to effective storage preservation and standstill protection is not a secret at all. Decades ago, plants employing best practices – also known as "BiC" facilities – began budgeting oil mist generators and associated equipment for dual use. Dual use refers to the oil mist apparatus being initially used for preserving the interior spaces/volumes of fluid machinery, certain valves, and all sorts and categories of other assets. Incoming machinery or machines delivered to job sites early or destinations not yet ready for installation, were placed on hardwood pallets located in outdoor storage plots, as shown in the aerial view of Fig. 6.8. The storage yard in Fig. 6.9 is ready to receive pallets with equipment; the yard begins to be populated in Fig. 6.10.

Fig. 6.9: Equipment storage preservation yard almost ready for incoming shipping crates. The oil mist unit can later be dismantled to become part of a permanent installation. Note manifold (circled) (source: Don Ehlert).

The machines in Fig. 6.10 and 6.11 are connected to the outdoor storage unit and flooded with oil mist at a pressure of 0.1–0.3 psi above that of the surrounding atmosphere.

It is worth noting that although the consoles or cabinets in Figs. 6.12 and 6.13 are designed and manufactured for permanent installation in a facility's process units, they can initially serve an entire outdoor storage yard.

Fig. 6.10: Outdoor storage yard. Note how plastic tubing is downward sloped to prevent pockets of oil from forming (source: T. F. Hudgins, Houston, TX).

Fig. 6.11: Close-up view of an outdoor storage yard with header, tubing, and actual machines filled with oil mist (source: Don Ehlert).

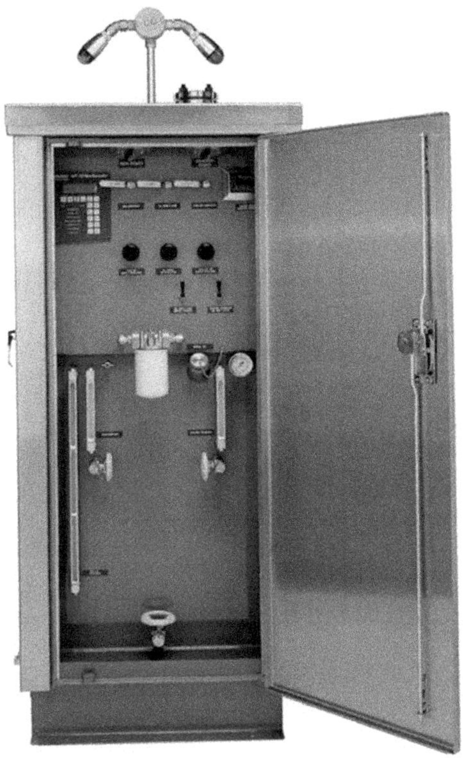

Fig. 6.12: Large cabinet, open view (source: Don Ehlert).

Once the storage plot and/or storage yard shown in Figs. 6.8 and 6.9 were no longer needed, the oil mist generating equipment ("OMG") was moved to its permanent location in the facility and expertly installed. The OMG then commenced many decades of successful, highly dependable service. It became the fully automated means of lubricating pumps, electric motors, and other equipment all within an operating radius of 600 ft (~180 m). It served both big and small machines, so long as these were designed for and incorporated, rolling element bearings. If a machine has rolling element bearings and these have been properly sized and installed, the machine can be lubricated with oil mist. Never once have we encountered an exception to these findings.

Again, and to emphasize: If a machine has rolling element bearings, properly applying oil mist from a very small, medium-sized, or extremely large plant-wide oil mist system will do the following:

1. Preserve and protect machines in outside storage for months and years. The oil mist enters the machine at a convenient location and fills the machine. A typical preservation pressure is 0.1 psi to 0.3 psi above ambient. Oil mist leaves at a small orifice located near the lowermost point in the casing or volume filled with oil mist.
2. Once the machine leaves its outdoor storage and is being installed on site, an oil mist generator that previously serviced the outdoor storage yard can be permanently installed in the process unit for the purpose of supply lubrication in

Fig. 6.13: Mid-size cabinet, open view (source: T. F. Hudgins, Houston, TX).

the form of atomized globules carried in clean air to a multiplicity of operating machines. The typical pressure upstream of an oil mist unit in plant-wide systems is 35 psig. The header pressure is 20–25″ of H_2O; once the oil mist has travelled into a bearing housing, the housing-interior pressure is between 0.1 and 0.3 psi.

3. Machines that are installed and serve a standby purpose are connected to the same plant-wide system. Whether a machine is operating or on standby, it will not be disconnected from an oil mist system.
4. When a fluid machine or its driver are taken from the field to the shop for any reason, a small cap with a drilled 1 mm (0.040″) diameter hole (an “orificed cap”) will be provided. The orifice keeps venting oil mist to atmosphere roughly equal in flowrate and pressure drop under as-operating conditions. There will thus be no need to make instrument adjustments in the oil mist supply cabinet.
5. Neither operating nor standby machines will contain liquid oil at any time. Once pure oil mist is used for lubrication or standstill protection, none of the usual maintenance-intensive components (constant level lubricators, oil rings, oil pick-up disks, breather vents, desiccant bowls, sludge and water observation cups, water drain petcocks, etc.) are needed any longer. These will be removed and discarded.

Decision makers often reject the recommendation to use oil mist preservation because their knowledge of the subject is based on what they may have heard in the distant past, or because mere lip service is being paid to the concepts of safety and reliability. Absurd statements regarding explosion hazards are occasionally voiced by individuals who have no understanding of explosion limits. A hydrocarbon mixture of one volume of oil in 200,000 volumes of air is several orders of magnitude too lean for such hazards. But it does take informed managers, technicians, and operators to make good decisions, and a well-informed value-adding employee can demonstrate valuable insight on the topic.

We are deliberately making the same point more than once in this book: Among the primary attributes of oil mist is its application while half of the machines in a plant are operating, and while half are in temporary standby mode. During the time when one half are being lubricated, the other half are filled with slightly pressurized oil mist and are thus protected from the ingress of atmospheric air. After 1 or at most 2 months of operation, the machines previously on temporary standby become the operating machines, and the previously operating machines are placed on temporary standby. The time-related split is roughly 50–50. There are compelling reasons why soundly managed process plant with two installed pumps per service will *NOT* run half of their pumps 99% of the time and the other half only 1% of the time [7, 10].

To re-state: Best-in-class facilities use oil mist for all machines, whether running or not running. Originally developed in the 1930s to continually lubricate pneumatic road construction machinery, oil mist technology is now a mature and widely used, highly cost-effective technology. It is easy to understand and visualize why this technology is also extensively and almost exclusively used in standstill (i.e., storage) equipment preservation. Both installed spares and standby machines benefit greatly from an oil mist blanket when not running. Without such a blanket, the vapor space in a bearing housing would fill with ambient air. When the ambient temperature goes up or down, the air is either expelled or new air is aspirated. Lube oil is thereby contaminated and bearing distress is initiated.

The use of oil mist is routine at BiC companies. A BiC company is one that excels in equipment reliability and financial profitability. At some refineries, thousands of process pumps and their electric motor drivers are lubricated and cost justification calculations are easy and accurate. Payback periods of around 1.6 years are the norm for plant-wide oil mist units at these companies. Their less profitable and more repair-focused counterparts are often disinclined to use oil mist because, as their reasoning goes, "now is now and the future is far out and quite uncertain." To which we can only add our heartfelt concurrence. However, by not using the best available technology today, or by continuing to make decisions based on anecdotes and opinions instead of facts, some companies drive their own future much deeper into uncertainty [8].

Companies that have no real interest in the future also lack staying power. All too often, their plants have been derisively called rust buckets, a term that aptly describes the condition of their neglected facilities in the years before their doors are shut forever. Still, despite the strictly short-term thinking practiced by these companies, they are correct in pointing out that it is indeed possible to get away with using less-effective lubrication and protection methods than oil mist. However, using less-effective protective measures makes economic sense for only very short time periods, and then only if generally favorable circumstances exist for a while.

Expounding on that reasoning we point out that whenever inexpensive, short-term methods are mistakenly applied for too long a time, there may be serious repercussions. Many times, oil mist preservation would have been a far better option to implement. Chances are that oil mist was not considered because the decision makers did not know its simplicity, low cost, or far-reaching technical merits. Why they do not know these attributes is difficult to understand and certainly not worth speculating about. With the foregoing in mind, the next subheadings, and even entire chapters were written to inform teachable decision makers.

6.3.3 Fifty years of oil mist lubrication and why oil mist excels

Since the late 1960s, oil mist has excelled as an unusually simple and highly dependable lube application method. Compared to traditional liquid oil sumps in pumps and the often ill-defined methods of applying grease to electric motor bearings, plant-wide oil mist systems are much more reliable and cost-effective than dealing with individually liquid oil or grease lubricated machines [46, 47]. As of late 2018, an estimated 160,000 process pumps and 50,000 electric motors were being lubricated with oil mist. Well over 3,000 plant-wide oil mist systems were then in highly successful use. At their respective final points of application, usually at bearing housings as shown in the pump illustrations (Figs. 6.14 and 6.15), the oil mist coalesces; it "plates out" on (or coats) all bearing surfaces. The rate at which oil mist coalesces is greatly accelerated the very instant an oil and air mixture contacts bearing elements in motion. Small globules of atomized oil are then knocked together and form larger droplets of oil; these are too heavy to remain suspended or carried along in air. As mentioned above, the globules of oil plate out on all bearing surfaces and provide an oil film that both lubricates and protects against corrosion [48, 49].

Especially notable is the protection of standstill equipment in facilities where both an operating main pump and an installed standby spare pump sit side by side. The same, of course, is true for many different machines equipped with rolling element bearings. Fans, small blowers, conveyors and other machines respond the same: Whenever bearings are not moving, fewer globules get knocked together and less oil will coalesce.

However, it only takes a small amount of oil to fully *preserve* the interior surfaces of a bearing's housing. Even more important is the fact that the oil mist pressure inside a bearing housing is perhaps 0.1–0.3 psi greater than the pressure of the sur-

Fig. 6.14: Midsize pumps and motors lubricated by oil mist (source: LSC, Houston, TX).

Fig. 6.15: Large pump on oil mist. Note branch line terminating in a manifold. Instead of using a return header, the owner is using a collecting pot from which coalesced oil is pumped back to the header (source: T.F.Hudgins, Houston, Texas).

rounding ambient air. Therefore, neither external water vapors nor particulates (e.g., dust and fine sand) can enter the bearing housing [50].

The capacity of an oil mist system depends on bearing size and distance from the oil mist generator (OMG). The OMG is a mixing valve or venturi (converging–diverging nozzle, Fig. 6.4), where lube oil and clean instrument air are brought together. There are no moving parts in the oil mist generator. Level sensors and mist density monitors associated with the oil reservoir and pipes are based on radar or infrared sensing technology; again, no moving parts.

From 5 to as many as 70 pumps and/or drivers can be connected to a single oil mist system. The technology is widely used in advanced oil refineries and chemical plants, in other words, best-in-class facilities. When electric motor drivers are served in addition to pumps, and if the oil mist is captured after traveling through equipment bearings (capture makes this a closed system), oil mist lubrication technology is cost justified and most often yields payback periods of considerably less than 2 years.

6.3.4 Primary advantages over conventional lubrication summarized

The following are the primary advantages of oil mist. It should be noted that none of these advantages have ever been disputed by knowledgeable professionals; issues only arose when managers asked to see detailed cost justifications.

Here are some additional details:

- Due to its very low maintenance requirements, pure oil mist lubrication, also called dry sump oil mist, represents both least risk and best available technology for lubricating equipment incorporating rolling element bearings.
- Generally speaking, process pump sets consisting of pumps up to 500 hp have been connected to oil mist systems on a plant-wide basis. However, many individual pumps and motors as large as 1,250 hp are known to operate with oil mist lubrication [32].
- Decades ago, industrial manufacturing giant Siemens showed that its rolling element-equipped electric motors, some rated as high as 3,000 kW, could operate on pure oil mist [15, 52].

6.3.5 Closed oil mist systems

Closed oil mist systems allow little or no oil to escape into the surrounding atmosphere. Textile machine manufacturers were among the first to use closed oil mist systems in machines other than process pumps. As early as 1950, a Swiss textile machinery manufacturer relied on closed loop oil mist systems in hundreds of machines. Bearings in the high-speed draw rolls of its tire cord production machines had to be kept cool and lubricated and only pure oil mist was up to the task.

The company, which has been in business since 1795, and a handful of others can attribute their longevity and success to intelligent decision-making. They have studied and embraced innovation long before the competition. No doubt, their early acceptance of oil mist lubrication has contributed to their admirable staying power. Likewise, recognizing and implementing superior lubrication methods plays an important part in pushing best-in-class oil refineries to the top. Understanding and optimizing lubrication play key roles in achieving decades of high asset reliability and corporate profitability.

Today, in 2022, most plant-wide systems are of the closed type. A closed system is depicted in Fig. 6.16. The header and branch piping systems were described earlier, and the combination is again depicted in Fig. 6.18. Most closed systems use a central collecting tank for the coalesced oil. A tank-mounted small blower creates a slight vacuum, thereby aspirating uncoalesced residual oil mist into its crinkled wire mesh. The remaining oil falls into the collecting tank and relatively clean air is exhausted into the atmosphere.

Fig. 6.16: A large "closed" oil mist system. Its dark-brown return header leads to a collecting tank and vacuum blower (source: Lubrication Systems, Inc., Houston, TX).

With the exception of the blower, closed systems incorporate no moving parts. Oil mist is applied without using sensitive oil rings. Oil rings are subject to abrasive wear and/or slowing down if the shaft system is not absolutely parallel. Using pure oil mist, service-intensive oil rings, constant level lubricators, and the traditional oil and grease refill labor requirements are eliminated [8, 19]. The labor requirement for pumps with oil mist has been estimated as one-tenth that of traditional lubrication.

In the years since about 1998, the earlier practice of allowing excess oil mist to escape into the atmosphere has been superseded by widespread use of closed systems. Closed systems avoid polluting the environment, with as much as 98% of the oil recovered for reuse. Some of the earliest closed systems have been in highly successful service since the late 1950s and now represent best available technology in all respects [40]. Plant-wide systems are almost completely maintenance-free and fully self-checking. Users do not have to rely on operators or maintenance workers to check and fill housings with oil or replace oil in bearing housings.

Better lubrication conditions exist because the oil coating on the bearings is always new. Lower bearing operating temperatures are routinely obtained. Reductions typically range from 10 to 20 °F or 6 to 13 °C. Power requirements are typically reduced by at least 1% and sometimes even 3% since bearings operate on a thin oil film instead of plowing through a drag-inducing pool of oil or grease [7, 8].

The extended mean time between failures (MTBF) benefits of oil mist over traditional liquid oil in sump lubrication have been well-documented and oil mist was included in the American Petroleum Institute's venerable API 610 pump standard, since 1989. In describing the basic oil mist generation process, it should be noted that the bearing housings, shown in Figs. 6.17 and 6.18 contain no liquid oil [8, 19]. Instead, an oil mist generator or OMG, Fig. 6.1, with no moving parts creates the oil mist in a central console, as shown in Figs. 6.2, 6.5, and 6.12, 6.13, and other illustrations throughout this book. A smaller unit, shown in Fig. 6.7, is sized to serve two to four motor driven pump sets.

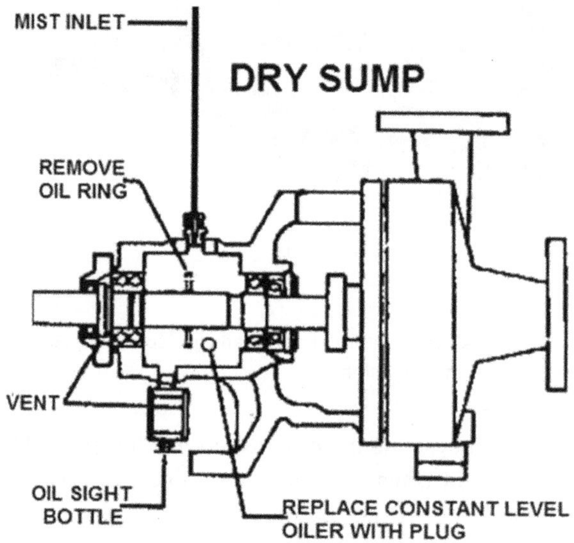

Fig. 6.17: In 1960s vintage pure oil mist (dry sump) applications, the oil mist entered at the top of the bearing's housing and escaped at several vent locations.

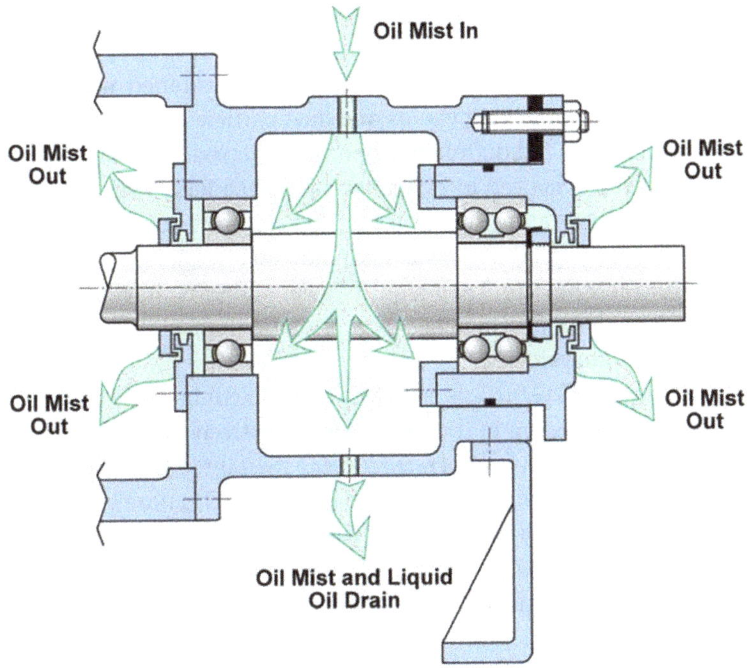

Fig. 6.18: Old-style (non-API-type) bearing housing with wasteful oil mist introduction at midpoint of bearing housing.

6.3.6 Operational parameters and simplified parts list

A typical console and its mist generator can lubricate a facility's process pumps within a radius of approximately 600 ft (~180 m). Although oil mist has occasionally been transported over distances as far as 1,000 ft (~300 m), a larger number of small oil globules will then collide and become too heavy for suspension in the carrier air. The large globules would drop out of the mist and slowly flow back in the slightly sloped header. With some oil now returning to the OMG, a leaner flow of oil mist would arrive at its final point of application. Therefore, using a 600 foot (~180 m) operating radius is highly conservative and will work in all conceivable climates.

There are no moving parts in oil mist generators. Synthetic lubricants with low pour points (–70 °F) and anti-wear hydraulic oil with an ISO viscosity grade (VG) of 68 are typically used in plant-wide oil mist systems. In moderate climates, a synthetic with ISO VG 32 would be feasible. While ISO VG 100 oils could be used, they may not remove as much heat from bearings and may, in some instances, not result in maximum bearing life.

A small electric heater is part of the OMG to maintain a constant mixing temperature which, in turn, ensures a consistent 200,000:1 air to oil (by volume) ratio.

Although a ratio deviation of ±10% is permitted, modern oil mist consoles are instrumented with density monitors to stay well within these limits.

Oil mist is a mixture of microscopic (<3 μm) oil droplets combined with clean instrument quality air. Plant air with water vapors and dust particles is unsuitable for oil mist. Consider it unacceptable and definitely responsible for reduced bearing life.

The oil mist is customarily conveyed in 2″ diameter pipe headers to virtually any location with equipment incorporating rolling element-style bearings. While flows and velocities can often be accommodated by smaller diameter headers, 2″ and even 4" headers are chosen for their rigidity. They require fewer supports, have low pressure drop, and optimally transport the oil mist at low pressure (typically 20″ of water column or 0.7 psi) and moderate velocity.

The velocity is kept below 7 ft/s (2.0–2.2 m/s) to reduce the likelihood or risk of globules becoming too large for suspension in the carrier air. Excessively large globules or droplets of oil would fall out prematurely or before reaching their intended destination. Near each process pump or electric motor, the oil mist passes through a nozzle or reclassifier, which is essentially a small-bore metering orifice. The mist velocity is thereby greatly increased. Additionally, a bearing in motion further promotes atomized droplets to collide and coalesce into larger liquid drops of oil. In both standby (i.e., installed spare) or permanently shut down equipment, oil mist serves as a protective blanket. Because its pressure in a bearing's housing is marginally higher than the surrounding atmosphere, oil mist prevents the entry of airborne dirt and moisture.

An entire system is depicted in Fig. 6.16. The OMG is where oil meets air. It is located inside an oil reservoir which, in turn, is designed to fit in an oil mist cabinet (lower right, inside the cabinet/console shown in Fig. 6.2. Piping consists of a delivery header in the foreground and a return oil collection header in the background of Fig. 6.16. Oil mist take-offs to and from process pump and motor bearing housings are connected to the top of their respective headers [50, 51]. In the very unlikely event that dust or other solids exist in the header, using top take-off locations allows only clean, uncontaminated, self-filtered oil mist to travel to an asset's bearing housing. Liquids, if they exist, stay near the bottom of the header and can be drained at convenient locations or, due to the slope or slight incline of the pipe header (0.1–1.0 degrees off horizontal), liquids and coalesced oil will flow back to the OMG.

6.3.7 Oil mist for plain bearings

Oil mist for plain bearings is feasible, but difficult to cost justify. Bearing housings of medium-sized steam turbines are sometimes equipped with plain Babbitt bearings that carry the radial load, while the rolling element thrust bearings located in the same bearing housing absorb the rotor's axial load. Plain bearings are traditionally, and usually optimally, lubricated with oil applied by circulating or liquid-in-sump systems, whereas rolling element bearings obtain optimal lubrication with either jet oil or oil mist (Fig. 5.5). However, questions are occasionally asked on the feasibility and desirability of oil mist being applied to both bearing styles.

Here's an actual example where a reliability engineer obtained an academic research article on the use of oil mist in plain bearings. According to the research article, in order for the oil mist to be applied to plain bearings, it has to be converted or reclassified to air-free liquid oil about 10–30 mm upstream of the plain bearing. This liquid oil is then allowed to flow into the plain (also called "sleeve") bearing. The reliability engineer was considering the use of oil mist in steam turbine bearing housings after first coalescing or "reclassifying" the mist into large oil droplets or even an oil spray. His rationale was supported by the existence of pipes that were originally intended for purge mist. (Purge mist is oil mist floating on top of a sump filled with liquid oil.)

The reliability engineer correctly surmised that it would be better to avoid, if possible, the use of an oil ring and liquid oil sump arrangement. This prompted him to inquire if he could pipe oil mist to within 10–30 mm proximity of the sleeve bearing and then convert the oil mist to liquid oil droplets. A separate oil mist line would enter the space between bearing protector seal and rolling element thrust bearing, thereby providing pure oil mist to the latter.

While conversion of mist to large liquid oil droplets had been done before and was well researched as a means of lubricating machine tools in the late 1950s, there are more reliable ways to lubricate plain bearings in small to midsized steam turbines. In plain bearings and with the use of conversion fittings, owners must consider all influencing factors around these bearings. Heat removal is one of the primary reasons for the oil, hence the flow rate of this reclassified liquid oil would have to equal the originally specified mass flow of oil. Ascertaining the needed equivalency may not be easy but would be required. The warmed-up liquid oil would have to be removed and recycled. In only a few highly specialized cases is the use of oil mist as a carrier for the later application and reclassification of oil mist for use in plain bearings cost justified. We have never encountered oil mist-lubricated sleeve bearings in the process industries.

6.3.8 Temperature limits for oil mist lubrication

Often, there are questions about what we call the "myth" of limiting product temperature when applying pure oil mist to rolling element bearings in process pumps. Some contractors use highly arbitrary guidelines that are easily refuted by many decades of field experience. Here is just one such example provided by a reliability professional:

> Our specification allows installing pure mist in pumps with product temperatures ranging from 120 to 250 °C, as long as a fan is installed on the pump shaft. Above 250 °C, the specification disallows pure mist and, instead, requires purge mist (i.e., wet sump). The very existence of a temperature limit set by certain design contractors is something I have difficulty understanding. At present, I am attempting to maximize the use of pure oil mist (i.e., dry sump) in the plant instead of the purge oil mist that had been widely used at our facility in the past.

We concurred and confirmed that the best refineries in the world use pure oil mist on just about any of their many thousands of rolling element bearings. Except for a few applications in small steam turbines or as a floating "blanket" in the space above the conventional oil level in bearing housings with sleeve or plain bearings, experts will not use purge mist in pumps. Pure mist is the gold standard and has been in use at many major refineries for well over 40 years. In view of decades of prior experience, the specification at this engineer's plant makes no sense. It will be easy to verify 40 years of good experience in pipe still bottoms pumps and other hot services with pumpage or product temperatures as high as 400 °C. Selecting proper lubricant application, using synthetic lubricants, observing proper bearing mounting tolerances and bearing installation techniques, we have never experienced housing temperatures in excess of 190 °F. While this temperature was found once in perhaps 1,000 bearings, it is decidedly too high to allow contact with an operator's hands or other body parts. Nevertheless, we hasten to add that modern rolling element bearings will work flawlessly and reach full life expectancy at operating temperatures up to 240 °F/115 °C.

It may be of interest that not so long ago there was a preference to move ambient air across the bearing's housing. In the days of using mineral oil for process pump bearing lubrication, forced air cooling was considered beneficial. Since then, however, synthetic lubricants have become widely available. In closed oil mist systems, the coalesced oil is collected after it has passed through the bearing toward a drain port near the low point of the bearing's housing. Moreover, best-in-class refineries have, for decades, applied pure oil mist in conjunction with appropriate synthetic lubricants (e.g., PAO or PAO/Diester-based mixtures) and no cooling of any kind is needed. Therefore, fans are no longer used. Also, water cooling is no longer needed on pump bearing housings and the associated water pipes were removed decades ago.

6.3.9 Hot bearings

If hot bearings (i.e., those exceeding 190 °F/88 °C) are encountered with oil mist, this is a sure sign of overloaded bearings, incorrectly installed bearings, incorrect bearing fits, incorrect lubricant viscosity, and so on. When bearings exceed 190 °F, it should bring to mind that all failures of rolling element bearings in process pumps are attributable to one of only four contributors: force, reactive environment, time, and temperature: FRETT (FRETT) [8, 10]. The FRETT method of identifying failure categories has been eminently successful since Knoxville/TN-based System Improvement Company made it the core of a highly acclaimed failure analysis course known as "Equifactor."

Still, a reliability engineer brought to light another point of view relating to hot bearings. He realized that certain feed pumps and pumps in hot pipestill bottoms service had bearing housings that were inches away from the pumps' fluid end. Moreover, the bearing housings were surrounded by air and he had given voice to the belief that his (European) refinery employer could decide to install pure oil mist regardless of fluid temperatures.

Indeed, this reliability engineer had correctly reasoned that pure oil mist is the right lubricant application even in his hot services because field measurements of the bearing's operating temperature had never exceeded 190 °F/88 °C. However, he advised that he wanted to submit to his managers assurances about pure mist working well in high temperature pumps. We complied with his wishes and reminded him of decades of relevant experience.

We have, on numerous occasions since 1975, confirmed that there are many hundreds of pumps in service with pumping temperatures of 740 °F/400 °C. Every one of these was lubricated with pure oil mist and some of these pumps have been in highly successful service for well over 40 years. Many of these were typically in operation for 6 or 7 years before being dismantled for precautionary inspections. Regrettably, that simple truth is occasionally disputed by people who have either no experience or who will, without hesitation, replace solid facts with totally unsupported opinions. They are, therefore, impeding the achievement of high reliability, as evidenced by repeat pump failures and demonstrably reduced profitability of their plants [8].

This is a good time to mention another fact: An API-style pump with rolling element bearings and a pumping temperature of 400 °C lubricated by pure oil mist will absolutely and unequivocally operate with greater probability of long bearing life than the same pump being operated with oil rings and a standard liquid oil sump. The only match for oil mist would be an auxiliary pump-around unit or the oil spray (i.e., jet oil) method alluded to earlier and again discussed later in this text.

In a 2017 case in the Middle East, it was claimed that a hot service pump with a 190 °F/88 °C bearing required liquid oil (!) lubrication. But, the folks at that location had likely overlooked an excessive interference fit between the bearing's inner rings and shafts. The pump at issue had a stainless-steel shaft; some stainless-steel shafts will thermally expand 17% more than typical ferrous tool steel shafts. The stainless-steel shaft aggravated the interference fit issue. The troubles at that location had nothing to do with oil mist, but rather were solely the result of human flaws, perhaps even ignorance or incompetence. Reading and acting on factual information from a 25 dollar book would have saved this refinery a small fortune.

6.3.10 Old-style open- and new-style closed-oil mist systems

In old-style open-oil mist systems (Figs. 6.16 and 6.17), the air–oil mixture fills the bearing housing, but not all the mist passes through the bearings. A portion of the coalesced droplets entering old-style open housings and systems take a straight top to bottom path through the bearing housing or enclosure. If the bottom drain port is left open or is connected to a small collector bottle, coalesced oil and some stray mist will exit and/or can be collected for disposal. For bearings to be properly lubricated in old-style open applications, the oil mist had to pass through the bearings and then escape to the environment at the two unsealed regions where the shaft typically protrudes through the bearing's housing.

Closed oil mist system technology differs, but Fig. 6.16 incorporates a collecting tank (shown at the far left) to which a return header system is connected. A small blower is provided at the top of the collecting tank and the suction effect of this small blower causes excess or stray oil mist to be pulled into the tank. Inserted in the blower is a coalescer maze. Coalesced oil droplets fall out and the oil can be reused. With highly efficient coalescers, almost entirely oil-free air is vented to the atmosphere.

However, efforts to simply provide effective bearing housing seals at the ends labeled "oil mist out" in Fig. 6.17 had unexpected consequences for inexperienced users. As mentioned earlier, oil mist works by causing small globules of lubricating oil to provide an oily coating on the bearing's components. In order to provide continuous oil replenishment on all bearing surfaces, the oil mist must flow and cannot be stagnant or dead-ended. When tight-sealing bearing protector seals are retrofitted to old-style configurations at the "oil mist out" locations shown in Fig. 6.18, there is no throughflow. Without oil mist flow, there is dead-ending and new oil mist cannot contact the bearing elements. Neither cooling nor lubricant replenishing can take place on the associated bearing surfaces [49].

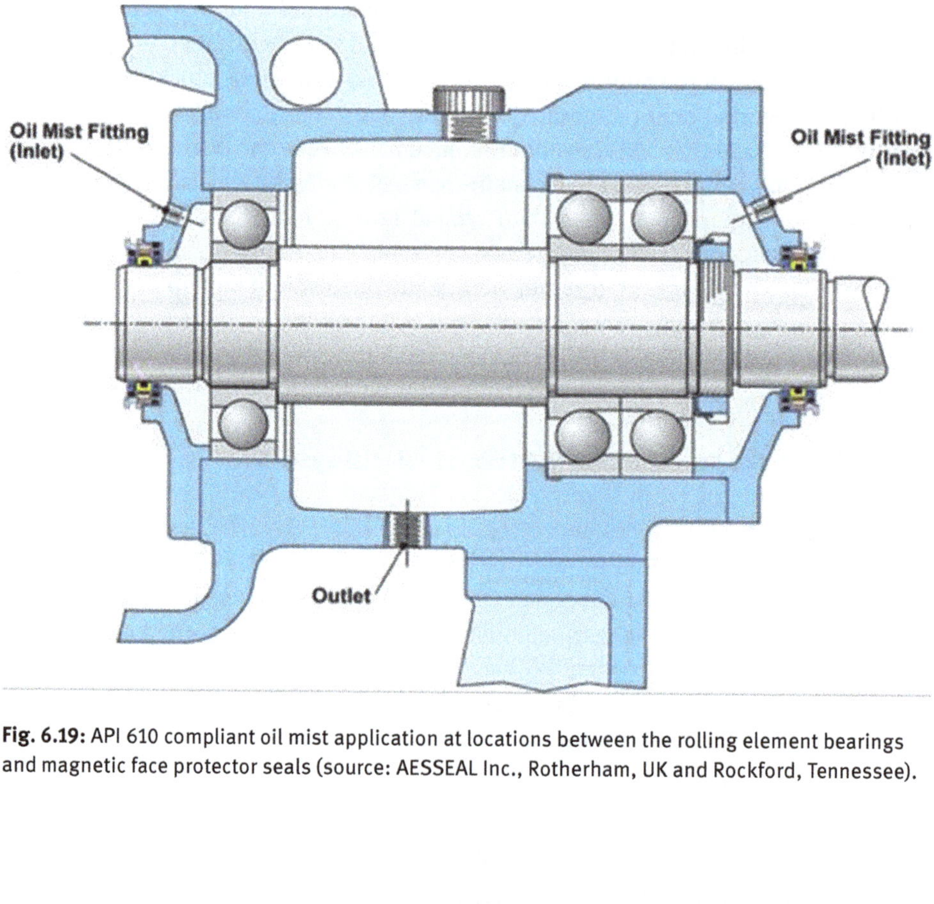

Fig. 6.19: API 610 compliant oil mist application at locations between the rolling element bearings and magnetic face protector seals (source: AESSEAL Inc., Rotherham, UK and Rockford, Tennessee).

Modern plants, which since about 1975 have always implemented the oil mist flow paths shown in Figs. 6.19 and 6.20, experience no such problems. Every single globule of oil arriving at the bearing housing's bottom drain port will have traveled through a bearing. Thus, every single globule has done its work and only clean, fresh oil has contacted the rolling elements. Because it is clean, oil leaving at the drain port can be reused after sending it through an elementary filter. Filtration is advocated to cover the remote possibility of a speck or sliver of bearing metal having made its way to the drain port. The replaceable filter can be as simple as a swatch of finely woven linen inserted in a pipe union.

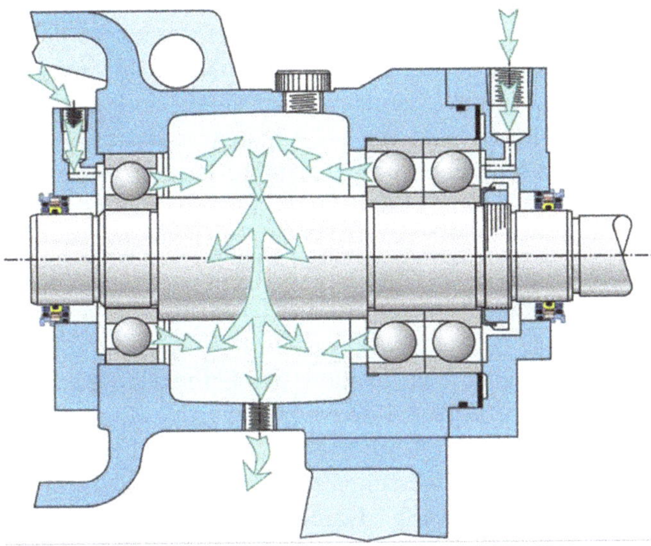

Fig. 6.20: API 610 compliant new-style oil mist application method. Note magnetic seals (source: AESSEAL Inc., Rotherham, UK and Rockford, TN).

Reliability leaders at BiC companies have fully implemented the most ideal routing of oil mist through bearings, as shown in Figs. 6.19 and 6.20. This routing became the standard configuration for BiC users around 1978. Note again that oil mist is introduced into the space between a modern bearing housing protector seal and the bearing. Any oil reaching the bottom center of the bearing's housing would have first cooled and lubricated the bearing. Also, having a single, centrally located exit location (a threaded port previously provided with a drain plug) makes it easy to collect the coalesced oil or residual oil mist. The system is thereby closed, and excess oil is ready for filtration and reuse. The oil mist inlet fittings are aimed at the bearing cages and the oil mist will overcome windage or fan effects caused by certain angular contact styles of bearings.

These or similar directed oil mist fittings [35] are routinely used by BiC companies; they become mandatory whenever shaft surface velocities exceed 2,000 fpm (ft/min).

A 2002 visit to eight petroleum refineries located in the US Gulf Coast region found that oil mist lubrication was the predominant method of lubricating pumps throughout the refining industry in the United States. In the same year, an equipment sales specialist who had worked as a refinery reliability engineer for over 30 years, estimated that oil mist was being used by 24 of the 30 refineries in the Beaumont-Port Arthur metropolitan area in southeast Texas. Furthermore, about 80% of the pumps in each facility were lubricated by oil mist systems. One US West Coast consulting engineer with considerable background as a refinery engineer estimated that about 55% of all oil refineries in the United States were using oil mist in 2018.

Several of these refineries have now employed closed oil mist systems for nearly five decades and are calling this application method an unqualified success. The refineries consider closed oil mist systems a competitive advantage and have fully endorsed the application routines illustrated in this book. Moreover, these users are doing their part toward achieving a cleaner environment while imparting reliability to their rotating equipment assets.

6.3.11 Quality of air needed for oil mist

Workers' health guidelines of many industrialized nations permit the relatively small amount of oil mist released from open systems. However, irrespective of prevailing, mandated, or government-legislated clean air requirements, an environmentally conscious user would not allow continual releases of oil mist into the atmosphere. Moreover, from a housekeeping viewpoint, it is clearly advantageous not to have smudges of oil on the ground near pumps and other equipment. In a refinery, oil smudges or oily rain runoff will reach a waste oil pit or API-separator, and it takes money to extract that oil before the water can be discharged. Therefore, an open oil mist system no longer represents best available technology, which is why proven closed oil mist systems and final application per Figs. 6.19 and 6.20 are greatly favored today [30].

Only clean, instrument-quality air should be used (ISO cleanliness grade 18 to 23, with grade 20 an easily achievable average). The air must be dry, with a dew point typically at −40 degrees. There have been highly misguided attempts to justify using relatively wet or dust-laden plant air supplies. Whenever someone tries to "save" in this manner, the perceived savings are quickly canceled by the incipient destruction of bearings, high maintenance costs and frequent downtime.

The issue merits emphasizing because taking chances with cheap lubricants and making foolish decisions devoid of common-sense reasoning cost the industry millions and gives rise to word-of-mouth anecdotes. Unscientific gossip travels faster than facts; mistakes are rarely admitted, and misinformation can give even sound technologies a bad name.

In closing this section, it should again be noted that oil mist is several orders of magnitude below the limit where it could explode. Sending oil mist into an open flame might make the flame burn brighter. Once the open flame is removed, oil mist will not continue to burn.

6.3.12 Modern bearing housing protector seals used with oil mist

The oil mist routing per Figs. 6.19 and 6.20, together with advanced face-type, fully sealing, magnetically closed bearing housing protector seals in Fig. 6.21 (left) or axially-moving O-rings (right) proves highly successful. Laser-etched face geometry was developed for virtually leak-free use on equipment lubricated by closed oil mist systems. Less expensive axial O-ring valve-equipped bearing housing seals, shown in Fig. 6.21 (right), became available soon after. They allowed barely perceptible seepage through the microlift gap and made closed systems possible. Predecessors to Fig. 6.21 are shown in Fig. 6.22 and a damaged O-ring from a long superseded "bearing isolator" is depicted in Fig. 6.23.

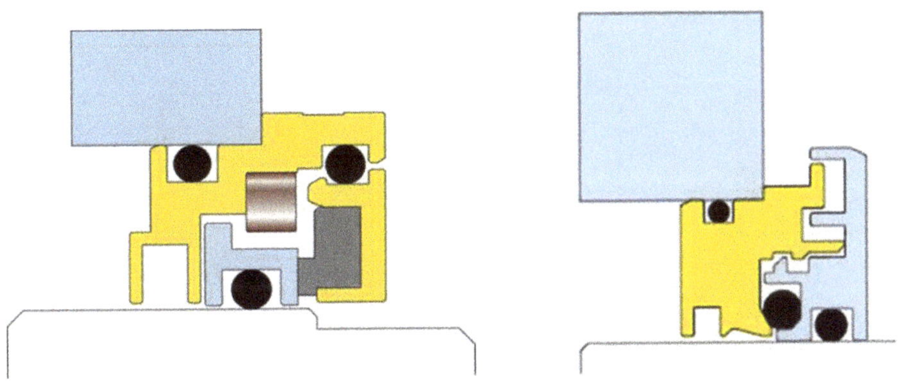

Fig. 6.21: Advanced magnetically closed bearing housing protector seal for oil mist containment (left) and a modern rotating labyrinth seal with axially moving O-ring (right) (source: AESSEAL Inc., Rotherham, UK and Rockford, Tennessee).

Forward-looking oil mist users have discontinued the old oil mist application per Figs. 6.17 and 6.18 since the mid-1970s and have since enjoyed decades of superior experience with the routing shown in Figs. 6.19 and 6.20. This routing was probably a first, initially specified in 1975 and fully implemented in 1978, in hundreds of pump sets at a grassroots petrochemical facility in the Southern United States. Unlike other plants that started with conventional lubrication and later converted to oil mist, the process pumps were purchased with neither oil rings nor constant level lubricators, two of the weakest or most failure-prone components of process pumps.

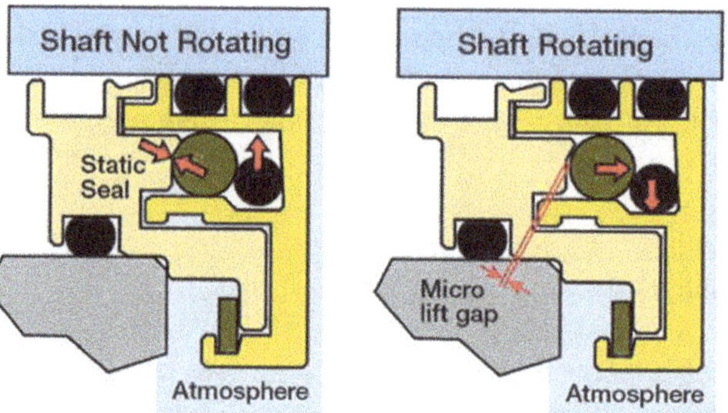

Fig. 6.22: Rotating, labyrinth-style bearing housing protector seals with large axially moving O-ring sealing off at standstill (source: AESSEAL Inc., Rotherham, UK and Rockford, TN).

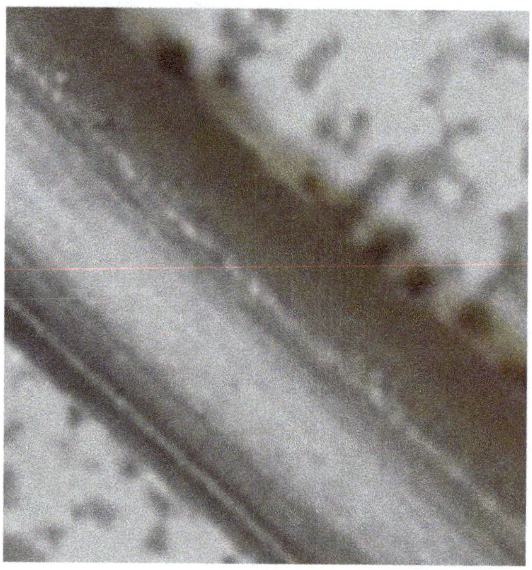

Fig. 6.23: Old-style "flying O-ring" bearing isolators were prone to be damaged.

In 1977, this leading facility began intelligent experimentation with high temperature (HT) mechanical seals. It started replacing segmented carbon ring gland fill in small steam turbines, using HT mechanical seals instead. Pure oil mist protected cooling tower fan gears, thrust bearing sets with dissimilar contact angles were installed, and many other equipment life-extending developments were implemented.

Using a leading plant's experience as one of its yardsticks, when the American Petroleum Institute released the eighth edition of API 610 in 2000, [21] introducing oil mist into the space between bearings and bearing housing protector seals had become the norm. As of 2021, the facility which in 1978 started up on oil mist (instead of converting from prior art) continues operating as one of the most reliable and profitable plants operated by its corporate owners.

The above-mentioned facility initially used mineral oils in its fourteen plant-wide oil mist units. The plant later began to cost justify and use superior ISO VG 68 mineral/synthetic hydrocarbon mixture formulations with pour points of −50 °C, or about −70 °F, and viscosity indices near 148. As the plant expanded, more oil mist systems were installed. In fact, it was determined (in 2016) that a vertical pump electric motor driver in the 30 hp range never had its bearings replaced in the 38 years since being commissioned in 1978.

Books and articles also describe the unparalleled success of oil mist lubrication for electric motors, a dozen or more are listed in our references. Contrary to unfounded and fully refuted opinions about oil mist attacking electric motor windings or oil mist being a fire hazard, motor interiors are not being degraded by oil mist lubrication. This was pointed out in [33] and published in 1977. In re-reading the article 42 years later, its author stated that he would not change a single word.

It is of real interest that even explosion-proof motors operate quite safely when lubricated and blanketed with the near-inert oil mist mixture at slightly higher than atmospheric pressure. In contrast, the same explosion-proof motor might experience intrusion of flammable and/or explosive gases if, say, a hydrogen-rich environment somehow found its way into the normally unpressurized interior of a motor. Many, if not all, of the unpressurized motors using one of the more traditional lubricant application methods are at considerably greater risk.

Experts investigate and then choose pure oil mist lubrication because it represents a fully proven and technically superior lube application method. Closed loop oil mist best protects industry's physical plants, as well as its environment, whether in full operating mode or standstill spare equipment. In BiC facilities, 97–99% of the lube oil is recovered and reused in closed oil mist systems. Closed oil mist systems emit little, if any, hydrocarbons into the surrounding atmosphere. For years, these systems have been preferred by responsible facilities. These include oil refineries, process plants, petrochemical plants and the many people who treasure clean air for every breathing thing.

Oil mist, supplied to the space between a bearing housing protector seal and its adjacent bearing, represents best practice lubrication.

Oil mist maintenance requirements are minimal.

Groups of operators and maintenance technicians can be fully instructed in oil mist technology in half-day tutorials.

In the 14-year period from 1978 until 1992 the total availability reached by 14 plant-wide oil mist systems exceeded 0.99995 at a South Texas facility. No other systems or lubrication methods have come close.

Bearing housing protector seals with axially moving active O-rings perform better than models with radially moving active O-rings.

Chapter 7
Oil mist history and reliability experience

7.1 Management digest

As of late 2018, an estimated 160,000 pumps and 52,000 electric motors were lubricated with oil mist. And as of 2021, two oil mist design and manufacturing companies located in Houston (Texas) are probably sharing 90% of the market for plant-wide oil mist systems. The technology is widely accepted by advanced and profitable oil refineries and petrochemical plants.

Probably 3,400 plant-wide oil mist systems are operating in close to 100 countries around the world, from Aruba to Zambia. An oil refinery in Venezuela used plant-wide oil mist systems as did refineries from Northern Canada to the southern tip of Argentina. One of the world's largest olefins plants used plant-wide oil mist system in the mid-1970s. Its hundreds of process pumps were purchased without constant level lubricators, oil rings, flinger disks, and other traditional components. Thousands of machines in India are lubricated by plant-wide oil mist systems as of 2021.

At an oil refinery in Aruba (Netherlands Antilles), in the late 1960s, oil mist proved by far the most successful lube application method on two 1,250 HP vertical electric motors [7.1]. In 1966, a major oil refinery in Venezuela was built with oil mist systems that duplicated several such systems in successful service in Louisiana.

7.2 Scope of overview

While the scope of this book does not favor repeating the hundreds of available narratives on the proven scientific basis of oil mist, there are oil mist handbooks, theses, and many other references dedicated to this topic [34, 35]. Similarly, reliability professionals can point to copious data showing that request for "proof of concept" installations are without technical merit. Those looking for proof of concept when seeking assistance in setting up field trials will find it in nearly 100 external references. All point to oil mist as a mature technology that has been used by a renowned bearing manufacturer since 1937 and by a textile machinery manufacturer as a closed indoor oil application since the mid-1950s. In the early 1960s, oil mist found its way into process industries.

7.3 Why oil mist is a mature technology

This gets back to an important point: Oil mist *lubrication* should always be mentioned together with the additionally relevant term of interest, oil mist *preservation* of equipment on standby and equipment in storage. Because oil mist inevitably preserves

https://doi.org/10.1515/9783110749441-007

standby equipment, the resulting reliability increase and maintenance cost avoidance deserve to be reflected in the cost justification, as should failure avoidance and the ensuing reduction in pump-related fires.

Needless to say, oil mist lubrication and preservation are even more easily justified in geographic regions with high humidity, regions with blowing sand, or parts of the world where lube automation is easier to accomplish than finding, training, and employing a truly dependable workforce. In northern climates, rotating machinery has occasionally been placed indoors in very large, heated buildings and lubrication concerns were given as the reasons. Oil mist lubrication does not present an issue in northern locations so long as a 600 W (maximum size) heater is included in the oil mist generator assembly.

After oil and air are brought together at or near an ideal mixing temperature of 100 ° F (about 38 °C) and a pressure of ~20″ (~500 mm) of H_2O, the oil mist moves from this header pressure to the much lower pressure that typically exists in a bearing housing. Oil mist will move in an unheated, uninsulated header and it will do so whether the external environment is at −45 °C in Canada or +45 °C, somewhere in North Africa or the Middle East. Visualize the oil mist to behave like cigarette smoke. Once we blow smoke into a 10-m long steel pipe, it will always arrive at the far end in the same form, that is, as smoke. This can be proven by simple science (except for what happens in a nuclear explosion, mass can neither be created nor destroyed) or by taking the experiment to the North Pole in late January or to the Sahara Desert in mid-August.

Additional benefits are derived from oil mist lubrication on equipment drivers. Indeed, every experienced plant applies oil mist to both its pump and electric motor bearings [52]. Recall that with oil mist, the air-oil mixture is routed through bearings in accordance with the guidelines set forth in the eighth and later editions of the API 610 standard (refer to Figs. 2.27 and 5.4). These editions and the latest 12th Edition (January 2021) depict the optimized flow-through method that has been in use at virtually every one of the world's most profitable refining facilities in the decades since 1960 [34, 35]. Moreover, hundreds of papers and articles have documented oil mist preservation.

Converting from grease to oil mist lubrication can be done with the equipment running or standing still. Conversion is especially straightforward in electric motors [43, 47]. Thousands of electric motors have been upgraded to oil mist lubrication by first removing the grease fitting from the top of the bearing housing. Next, the drain plug is removed from the bottom of the bearing housing and a vacuum hose applied for about 30 s to remove much of the grease. A 1 mm diameter vent hole is drilled through the drain plug, which is then re-inserted in its former location. Finally, a reclassifier fitting (orifice) and associated tubing are connected to the top of the bearing housing in the exact thread from which the grease fitting, often called a "Zerk fitting", had been removed. The drain can also be connected to a closed system.

7.3.1 Few maintenance tasks with oil mist

At International Pump Users Conferences, engineers and technical staff from user companies, including some of the largest multinational refiners and petrochemical companies, freely shared their highly favorable experiences. Of course, reliability professionals at these facilities were in the business of keeping the plants running. At the same time, they were tasked with finding cost-effective ways of extending and optimizing the equipment uptime. Their willingness to share their findings in papers, articles, conferences, and technical society meetings, over the years has been of help in developing best practices for oil mist. There has been a unanimous corroboration that oil mist systems are completely self-checking and virtually require no maintenance. A Texas-based plant with hundreds of API-style process pumps and numerous general purpose machines contracted a competent maintenance worker to spend a single day per month at this facility. The worker reviews the status of the 14 systems at that plant and visits the site one day each month. Maintenance requirements are truly minimal.

Optimizing uptime does not mean adding maintenance cost and, in fact, implies that far fewer maintenance interventions will be needed. The final outcome and ultimate test of a best-in-class (BiC) facility has been, and will continue to be, safe operation at the lowest possible lifecycle cost of all assets. In many places, oil mist storage preservation and oil mist lubrication have assisted in meeting or exceeding the highest safety and lowest possible lifecycle cost expectations. This explains why the author has again placed considerable emphasis on oil mist preservation, although other old style equipment protection methods are also described in detail later.

7.4 Relating oil mist experiences

Around 1998, the CEO and the majority shareholder of a prominent lubrication provider reported that plant-wide oil mist lubrication had been installed in over a hundred oil refineries and chemical plants in dozens of countries. That report came from just one provider company, although several such provider and service companies existed then, and today in 2021.

While we are not permitted to publish the names and locations of plants that supplied relevant data, we can share the benefits calculated by some of these oil mist users. It should be noted that these data include numbers that one can use in calculating cost justification and payback. Keep in mind that today, over 22 years later, oil mist providers may see fit to be much more specific in confidential discussions. We know that oil mist applications exist on all inhabited continents and large orders for advanced oil mist units are being fulfilled even as this book goes to print, in late 2021. That said, here is what we can freely share.

- Satisfied users include major multinational companies as well as small facilities.
- The investment made by these plants generated attractive returns and short payout periods, based on improved equipment reliability and reduced maintenance cost.

The known areas of improvement and achievements with pure oil mist lubrication include:

1. Reduced pump and electric motor bearing failures:
   - An 80–90% reduction in pump bearing failures is typical.
   - Electric motor bearing failures are often lowered by over 90%.
   - Competent oil mist suppliers can provide data on:
     a. bearing failures at a major refinery in Thailand;
     b. a California refinery sharing its electric motor failure history;
     c. bearing failure histories at a major olefins plant.
2. Reduced number of mechanical seal failure events:
   - Reduction of seal failures is in the range of 30–50%.
   - One user reported that the average mechanical seal life doubled to 8 years.
   - Examples included:
     a. bearing and seal experience of an oil mist user in California;
     b. seal life comparison from an offshore facility.
3. Reduced failures rates of specialty equipment:
   - Oil mist application has shown excellent results in a variety of other equipment applications.
   - Rotary lobe blowers, chemical mixers and cooling tower fan gearboxes are examples of more specialized applications with large payouts.
   - Examples include:
     a. polymer processing equipment failure history;
     b. applicable experience with rotary blowers;
     c. highly favorable refinery cooling tower gearbox history.
4. Results expressed as mean time between repairs/failures (MTBR/MTBF) showed significant improvement for pumps, drivers and other equipment:
   - One user reports improvement from 3 years (before oil mist) to 9 years (after oil mist).
   - Another user went from 4 years (before oil mist) to almost 8 years (after oil mist was introduced).
   - Detailed examples are available for:
     a. a refinery in a Pacific Rim country that published seal life comparisons and highlighted how bearing issues lead to seal failures;
     b. highly favorable pump MTBR experience;
     c. similarly favorable small steam turbine MTBR experience.
5. Disclosure of pump maintenance costs, showing significant reductions. Percentages are given; for example:
   - One user reports a 40% reduction in all work orders for pump maintenance.

   - Others reported pump repair costs reduced by 60–80%. These included:
     a. an asphalt plant in the United States;
     b. experiences at several refineries;
     c. dollar cost reduction numbers provided by one refinery.
6. Operations manpower to carry out lubrication tasks was reduced. Examples include:
   - A user reporting a 47% reduction in hours needed to complete lubrication-related tasks.
   - User feedback from a Pacific Rim country supported the data.
7. Lubricant consumption was reduced:
   - A 40% typical reduction due to more efficient application of lubricant.
   - One user reduced consumption by 70% by applying recommended oil recovery steps.
   - A comparison of oil usage in several affiliated Pacific Rim refineries is available from one of the two major suppliers with world-wide presence.
8. Reduced energy consumption is a fact:
   a. At a minimum, a 1–2% lower energy use was demonstrated in several controlled tests. Some tests by a major bearing manufacturer demonstrated a 3.6% efficiency gain.
   b. A South American energy consumption study has been published.
9. Eliminated lost production incidents:
   a. A specialty polymer producer estimated a 7–8% run time improvement.
   b. A refiner eliminated costs from lost production incidents on a crude oil unit.

The overall economic results from five refining applications have been published. The results are:

A. Western United States Refinery:
a. applied oil mist to crude unit, fluid catalytic cracking unit (FCCUs), and steam boiler area in 1999;
b. experienced sharp reduction in pump maintenance costs;
c. eliminated lost production incidents on crude unit;
d. discounted cash flow (DCF) returns exceed 200%;
e. several of the above results were achieved in less than 1 year.

B. United States Southern Great Plains State Asphalt Plant:
a. one system serving an entire plant was installed in 1997;
b. pump repair costs dropped 72%;
c. DCF return of 150%;
d. payback was obtained in less than 1 year.

C. Overseas refinery:
a. installed systems throughout a plant in the mid-1990s;

b. compared performance with sister plant (without oil mist);
c. results include doubling of MTBR for pumps and seals, cutting operating manpower in half, and reducing lubricant consumption;
d. estimated DCF return for converting the other refinery to oil mist is 54%;
e. estimated payback is only one and nine tenths years.

D. United States Mid-Coast Refinery:
a. two systems installed in 1996 on crude processing units;
b. pump bearing repair costs dropped 88%;
c. DCF return of 70% based only on lower repair costs;
d. payback was achieved in one and five tenths years.

E. Southern United States Refinery:
a. three systems installed in 1989;
b. pump repair costs reduced by 65–70%;
c. DCF returns of 75% achieved, based only on pump repair savings;
d. payback in one and five tenths years.

Similarly, the overall economic results for three petrochemical applications are available. They show the following results:
A. Specialty Polymer Plant in a US Western State:
a. failure rate on rotating equipment was about once every 6 months before oil mist;
b. failure rate dropped 98% after mist was applied;
c. plant availability to manufacture polymer increased 5–7%;
d. DCF return, without including increased production, exceeded 400%;
e. payout in less than 6 months.

B. Commodity Polymer Plant in US Mid-South Region:
a. high rate of rotary lobe blower failures prompted oil mist investment;
b. blower maintenance costs reduced by 90% within 2 years;
c. resulting DCF return of 45% and a payback period of 2 years.

C. Central U.S. Gulf Coast Olefins Plant:
a. compared pump failures between plant built in early 1980s, with oil mist, to one built 10 years earlier using conventional lubrication;
b. pump bearing failures 90% lower in oil mist lubricated plant;
c. DCF return, based only on fewer bearing failures, is 75%;
d. incremental cost of oil mist implementation yielded payback in one and one-half years.

In early 2005, a Saudi Arabian refinery reported accurate tracking of 1,400 process pumps lubricated by pure oil mist. In the previous year, there were no bearing fail-

Fig. 7.4: Close-up of a cooling tower fan gearbox equipped with oil mist (source: Don Ehlert).

Key improvements have been listed below; they combine to spell success. Using pure oil mist for bearing lubrication, and blanketing (with oil mist) the previously moisture-laden space above the liquid gear oil resulted in:

1. Elimination of a minimum of two gearbox failures each year.
2. Safety improvement; entry permit for two mechanics no longer needed.
3. No need for locking out operations-involved fan drive motors; scaffolding construction inside the cooling tower cells discontinued.
4. Environmental improvement; no oil spill risk with draining and filling gearbox while fans in operation.
5. Eliminated the need of a mobile crane for removal of the defective gearbox and reinstallation of the repaired gearbox. Watch personnel on the ground no longer needed.
6. Electric motor now pure mist lubricated, no more over-greasing of bearings.
7. Winter windmilling in reverse was an issue. Although not curing the windmilling, oil mist now coats all internals with clean fresh oil. Failure risk is reduced to an extent.
8. Atomized oil, suspended in air, exists at 0.1–0.3 psi above ambient. Therefore, water vapors cannot enter and greatly improved gearbox reliability is now a fact.

7.5.2 Case history: rapid payback from modern oil mist systems at an oil refinery in Texas

A large West Texas oil refinery applied dry sump (pure) oil mist lubrication to all centrifugal pumps throughout its crude processing units and catalytic crackers. In addition, purge oil mist lubrication, or *wet sump*, was applied to the small steam turbines and gearboxes. The project was started during 2000 and was completed during 2001. In total, 6 oil mist lubrication systems were installed to serve 130 pumps, 31 steam turbines, 26 fans and blowers, and 25 gearboxes in 8 processing units and 5 cooling towers.

This refinery captured several significant benefits from the application of modern oil mist systems. Here are the seven main points that the refinery has recently shared with their Houston-based supplier:

1. The number of bearing failures dropped from 18 in 1999 to 4 in 2001. This 80% reduction is typical and consistent with the experience documented by others over the past three decades,
2. Mechanical seal-related failures during the same period declined from about 77 to 38. The refinery attributes this 50% seal failure reduction to the fact that seals last longer when bearings run smoothly and do not cause internal component alignment problems. Reductions in bearing and seal failures, together, have extended the pump MTBR (mean time between repairs) from 35 to 75 months since applying oil mist.
3. The refinery also observed that, in the few instances involving failures of equipment lubricated with oil mist, the average cost of a repair was about 15% lower than what it was before using oil mist. The failures, on average, have resulted in less damage to the equipment and have thus reduced the need to extensively rebuild pumps.
4. The most impressive part of this story is that the repair cost, resulting from unplanned mechanical events, decreased from about $1,400,000 in 2000 to $700,000 in 2002, a reduction of 50%. This cost reduction alone provides a strong return on the investment in oil mist.
5. In addition to the sizeable direct savings, this refinery reports some other interesting statistics. The number of pump mechanics in the plant has been reduced by six. The total number of work orders of all types has come down by about 10%. In addition, the time spent to perform routine preventive maintenance on pumps was reduced from 4 h per pump (before oil mist) to only 1 h per pump per year (after effecting the oil mist conversions). This is because it is no longer necessary to periodically drain and replenish oil in the bearing housings.
6. The impact of oil mist on turbine performance was drastic. In this project, 31 small steam turbines were fitted with purge oil mist, whereby the oil mist serves to exclude atmospheric air from the bearing housing. Oil mist fills the housing interior, but liquid oil is used to supply the bearings with lubricant by traditional

means of application. A steady upward trend in MTBR on small steam turbines has been observed – from around 50 months in late 2001 to 150 months in the third quarter of 2003. This strongly suggests that purge-misting steam turbine bearings prevents the ingestion of airborne dust and atmospheric moisture contamination.

7. There is more interesting news to this success story: This oil refinery, during the peak gasoline production season each year, operates the spare pumps on the crude processing units in parallel with the main pumps in order to maximize its crude oil processing capability. The refinery was experiencing an average of five failures each year on the pumps running in parallel. This forced a reduction in the daily throughput of crude oil until the defective pump could be restored to service. Following the application of oil mist, there have been no failures of these critical pumps. The estimated annual credit for eliminating lost production is $500,000.

When these economic credits, for eliminating lost production, are added to the large savings in maintenance costs, the result is a return on investment of well over 100% and recovery of the initial investment in just over eight months. This application re-emphasizes why we consider oil mist for modern process plants.

7.5.3 Updates always confirm earlier findings

A world-scale oil refinery engages in plans for expanding throughput in process units or becomes the licensor for some of its process technologies. Based on custom and experience, the choice of properly designed and installed oil mist is not an issue with this refinery. Regarding a process unit commissioned circa 2001, the refinery has had all its pumps running on oil mist for 14 years and had not had a pump out for bearing maintenance by the report date (late 2015). They considered it, after 14 years, an infinite MTBR (mean time between repairs). The refinery is processing relatively clean gasoline products and its pumps were properly sized for the various services. Sister units using oil mist have had a similarly long MTBR. Operations teams and reliability groups believe in the results achieved with oil mist, along with proper sizing and proper installation of process pumps. The statistical evidence is very supportive of the choices made since the early 1970s when oil mist was first demonstrated to this facility.

A few years later, in 2018, this refinery reported the pump MTBR in quotation marks: "Our mean-time-between- repairs is now somewhere between 114 and 120 months, depending on a lot of things, but we consider these numbers close enough. In any event, we still have not had a lubrication-related bearing failure since 1999 that is worth mentioning. Note that as of 2018 we have ACHE (air-cooled heat exchanger) fan bearings that have not been touched since the fans were installed in 2004. Moreover, the same impeccable results were obtained with the pillow block bearings on the FD (forced draft) and ID (induced draft) fans at our refinery, which, as you know, is the largest on this continent."

7.5.4 Fewer shutdowns on record

According to the primary providers of plant-wide oil mist systems, only three shutdown incidents are known to have occurred on oil mist systems since 1982. An estimated 3,400 plant-wide systems have been in successful service for decades. In 1978, the author was involved in the startup of a world-scale grassroots olefins production facility, with 14 oil mist units, initially. The plant was later expanded to over twice its original nameplate size and several additional oil mist units are in service today, in 2021.

In the 8 years, from 1978 to 1986, a single "event" was reported when 1 of the 14 units annunciated an alarm [19]. The alarm panel light flashed because someone had forgotten to replenish the oil in a local oil supply tank. No pump bearings were lost in the incident. Oil wetted bearings remain oil wetted for several hours before their oil film is gone and the bearings begin to overheat. Modern oil mist units combine sensors and annunciators that give ample warning in the highly unlikely event of a malfunction. Recall that the only moving part in a plant-wide oil mist system is a solenoid that admits oil from an on-site tank (see page 168) to flow into a small tank inside the oil mist cabinet. A level-sensor signals the solenoid to open and close. Nothing else moves or rotates between a large oil storage tank and the bearing housings of machines lubricated with oil mist. Even the risk-inducing oil rings are gone from oil mist-lubricated equipment.

The first recorded systems interruption involving a modern plant-wide oil mist system occurred at a US Gulf Coast facility around 1982. At that time, a thorough analysis traced the failure to pipe shavings in the five-gallon capacity misting chamber reservoir. Ferrous debris became attached to a magnetic level switch, preventing it from activating a solenoid. Uninhibited solenoid movement would have allowed lubricant from a bulk oil holding tank to replenish the much smaller chamber reservoir. When the small reservoir was depleted, none of the connected equipment received oil mist. The bearings ran dry but did so without incident or bearing failure.

As an aside, the question on how long one can run standard bearings without lubrication was answered in the early to mid-1960s. It was then known that bearings coated with oil in a horizontally installed shaft systems can operate for about eight hours after discontinuing oil mist flow. Rigorous test protocols were developed by engineers at the Dow Chemical plant in Freeport, Texas and the results were reported and explained by Dow Chemical's Allen Clapp and Fred Wilcox [31]. Their test-based findings were again corroborated in full-fledged academic research conducted by Abdus Shamim, in pursuit of a doctorate at the Texas A&M University [35]. The observations by Clapp and Wilcox indicated that a small pool of oil will collect in the 5–7 o'clock segment of the contoured raceway in a bearing's outer ring. Oil mist lubricated pumps can safely stay in service for approximately 8 h before this small pool of oil is depleted. Given that there is ample supervisory instrumentation to indicate deviations, no modern oil mist system has ever encountered unavailability exceeding eight hours in the decades from 1972 until 2021.

The second unavailability event developed at an oil refinery in Enid, Oklahoma, where a single oil mist console was serving two adjacent process units. When the process unit where the oil mist generator (OMG) was located had to be shut down in preparation for a scheduled maintenance and repair downtime, the OMG cabinet switching valve was inadvertently closed. A day or so later, the adjoining process unit experienced a pump failure. It was immediately realized that there had been no oil mist supplied for at least twenty-four hours. The oil mist supply was then quickly restored, and no other bearing failures were experienced on any of the connected pumps. The cause of the failure was clearly human error; it could have been averted with a simple advisory note posted at the appropriate switch or valve.

A third incident report relates to a Texas Gulf Coast oil refinery where the owner-purchaser had opted not to include an automatic fill option on the OMG console. A manual reservoir refill line connected a bulk storage tank to the small, 10 L oil reservoir located inside the main oil mist console. An operator decided to crack open the needle valve in the refill line, expecting it to slowly maintain the oil level in the small reservoir.

In time, the entire piping distribution system was filled with liquid oil and the liquid oil had, in fact, displaced the atomized oil mist. About 12 pump bearing housings and their respective motor driver bearings were filled with oil. Although much oil was wasted, there were no equipment failures in this incident.

It is very difficult to remove the human element from a plant. All checklists, procedures, signal lights, bells, whistles, tablets, and laptops are of no use if a person decides not to pay attention to these devices. But here is the good news: Modern oil mist systems are provided with suitable supervisory instrumentation. As a result, there are no known reports of any system being disrupted for more than 2 h, in the years from 2000 to 2018. In 1998, overall reliability and availability were calculated by various observers as 99.99962% (lowest) and are currently thought to have reached 99.999997% (highest). The various availabilities recorded and reported for plant-wide oil mist systems have never been approached by any other lubrication method [7, 43].

7.5.5 What can shut down an oil mist system?

Considering the aforementioned three incidents, one has to wonder what else could cause an oil mist unit or system to shut down. Well, running a forklift into the 2″ oil mist header would shut the system down. It's a rather remote possibility, since such an incident has never been reported on any of the estimated 3,400 plant-wide oil mist systems now in service all over the world (August 2021). But, if it did happen, it would certainly take fewer than 8 h to repair a header which, in operation, is filled with low-pressure, nonflammable oil mist.

Actually, the main concern should not be damage to an oil mist header, but rather the hypothetical forklift operator conceivably driving his vehicle into a not-so-hypothetical process pipe. That pipe could be full of a pressurized or non-pressurized, highly flammable or perhaps explosive hydrocarbon product, a toxic insecticide, or a carcinogenic herbicide. These concerns, if they existed, would deserve to be addressed long before anyone would worry about oil mist.

7.5.6 Installed spare modules (mixing chamber reservoir) options

Since very little can shut down an oil mist system, the most profitable and reliable plants have found it neither cost-justified nor risk-justified to install oil mist systems with fully connected, ready to go spare backups. Nevertheless, and as mentioned earlier, some plants have occasionally asked oil mist systems vendors to propose and provide 100% redundancy. In those facilities, a backup or auxiliary oil mist system can be placed in operation at a moment's notice. Complete switchover from the main unit to a full backup takes thirty to sixty seconds. The switchover procedure calls for a one-quarter turn of the handle of a ball valve. A two- or three-sentence procedure sheet is posted on the inside of oil mist cabinet doors. The sheet explains what to do in the highly unlikely event of such switchovers ever becoming necessary.

Approximately 10% of plants with oil mist are using an installed spare oil mist module. The module consists of the combined components, "H" and "I," shown earlier in Fig. 6.2, mounted on the back of the console. The needed piping and quarter-turn ball valve connect to the same oil tank and pressure-regulated instrument air supply as the console. However, in view of the unparalleled reliability and availability of modern oil mist systems, the remaining 90% of users have seen no need to either purchase parallel units or mount spare modules on their cabinets or consoles.

7.5.7 Thoughtful layout saves money

The author was very impressed by a motivated EPC (engineering/procurement/construction contractor) working closely with a knowledgeable owner's SME (subject matter expert). The team realized that at least one major supplier of both oil mist technologies offered oil mist generator (OMG) heads with overlapping BI (bearing-inch) capacities. Their plant layout included a process unit "A," requiring an actual 472 BI, and a process unit "B," requiring an actual 352 BI. They specified, purchased, and installed two identical 500 BI units at the location where "A" and "B" met. Having ascertained that the two OMGs could actually generate as much as 800 BI each, the team devised cross-over valving, which allowed one of the two OMGs to act as the spare unit for the other. Oil mist technology and good brains make harmonious teams.

7.6 Warehoused spares

One or two spare modules are kept in the storage or warehouse facility of modern plants. These spares are configured and instrumented for dual purpose: first, as spares identical to the module mounted inside the oil mist cabinet and, second, as spares that could be set up for temporary or long-term use, perhaps next to a machine with risky or known to be unreliable lube supply. It should be noted that such warehouse spares must be designed and configured to serve equally well as replacements for the OEM-supplied oil mist module and as upgrade modules for existing machines. Such upgrading may become necessary and unavoidable for machines, which, after failing repeatedly, must be retrofitted with reclassifiers (application fittings) for oil mist.

The details of a modification can vary but are typically limited to installing a different size air–oil mixing component. Two configurations predominate. Venturi styles are a miniature version of the well-known ejectors and eductors we use to keep steam condensers at a partial vacuum. The second style is called a vortex generator. Either style is acceptable for mounting in the oil mist cabinet or the spare storehouse module (recall seeing both in Fig. 6.1).

Only the venturi or vortex generator (the OMG) may have to be changed to suit the flow range needed for the number of bearing housings in the machines served by a module. This change can usually be done in less than an hour. However, on some pumps that are retroactively converted to oil, the conversion work may take a few hours.

Figures 5.4 and 5.5 are typical of adding a directed reclassifier passage to each of the two bearing housing end caps. Reclassifiers (application fittings) can also be external and threaded into the top of a bearing housing, between the end cap and bearing. Alternatively, reclassifiers can be threaded into manifolds located at the ends of the so-called drops (branch pipes from the top of an oil mist header to individual equipment items). Directed reclassifiers will be needed if the shaft-surface velocities exceed 2,000 fpm (~10 m/s). This rule of thumb is based on the fact that the windage (fan effect) created by some angular contact bearings must be overcome by aiming (directing) the oil mist flow.

7.7 Oil mist is the ultimate filter

Oil mist is generated in a venturi or vortex generator, where oil and instrument-quality air are mixed in an ultra-lean ratio of 200,000 volumes of air to one volume of oil. Depending on the venturi size, the mixing ratio at the throat of the venturi may actually be somewhat less lean because large drops of oil are lifted from the reservoir into the air stream. However, large drops are too heavy to remain suspended in the "carrier air." They may also strike an impingement baffle and fall back into the reservoir. At 20–40″ of H_2O pressure, the 200,000:1 mixture will rise in a vertical pipe and float into a slightly sloped 2" or 4″ header.

The header is sloped because thousands of tiny oil globules will collide and grow too large and heavy to stay afloat. They will have coalesced into drops, which, by virtue of the header's slope, will slowly flow back into the mixing reservoir. While it would be quite feasible to use much smaller diameter headers to accommodate whatever flow "X" cfm (cubic feet per minute) might be needed to cool and lubricate a particular number of bearings having such-and-such diameters, a 4″ header is relatively rigid. It will not sag very much between supports, whereas a smaller header would require a few more supports to limit the pipe sag.

Pipe branches, called "drops," are either threaded or welded to the top of the header; 1″ diameter "drops" are typical. Because contaminants – if they exist – will stay behind in the header, the oil mist making it to the "drops" is ultra clean. Oil mist, making it from the rising part of the drops to the destination bearing housings, is free of solid contaminants and an oil mist system has thus become the ultimate filter [51].

Bearings run cooler with oil mist (typically 15–18 °F, or 8–10 °C). Indeed, the ability of the oil mist to preserve physical assets goes much farther than the generally well publicized attributes of continuously applying lubricant and blanketing to both, operating and on-standby pumps and electric motors [52, 53]. Dead-ended oil mist branch lines that terminate over the equipment can provide early warning of developing pump and motor fires, as will be described next.

7.8 Why oil mist terminations with low melting point alloys can be fire monitors

Fire abatement sprinkler systems are used in virtually all countries on earth. They consist of water-filled headers or branch pipes that terminate in a low melting point alloy fuse plug. As heat rising from a fire in the general vicinity of the pipe termination causes the fuse plug to melt, water rains down on the fire.

It is possible to provide oil mist drops or branch lines, with fuse plug terminations, directly over the mechanical seals or bearing housings of pumps and other equipment. Depending on the branch line size and fuse plug diameter, a relatively large volume of oil mist will escape and cause the header pressure to decay almost instantly. Pressures below the normal operating range setting would trigger visible and audible alarms, within seconds of heat causing the overhead fuse plug to melt. Once the fire is extinguished, a new fuse plug would be inserted, and the header pressure restored to normal. The alarm would be cleared, and none of the surrounding process pumps would experience bearing distress. As mentioned earlier, the lubricant that had previously coalesced or plated out in rolling element bearings would be sufficient to safely run without oil mist (or with a reduced oil mist supply) for up to eight hours [31].Uptime (equipment availability) would improve [54].

7.9 Using and supervising your own workforces to implement large-scale oil mist systems

It is entirely possible to involve many of your own staffers or semi-permanent contract work forces in implementing, operating, and maintaining large scale, plant-wide oil mist systems. Whether it is cost-effective to do so depends on timing, skills, management support, motivation, and so forth.

Here are some of the steps you can take to track costs, record progress, and ensure full technology transfer from an experienced firm to you:

1. Designate a "quasi owner" and involve this employee in the project. Engage the services of an experienced firm. Use the parts list and piping procedures (sloped header, connections, chip-free threads, etc.) provided by this knowledgeable oil mist systems design and installation contractor.
2. Understand both installation details and work processes. Understand the material to be supplied and how it will be handled.
3. Commence your involvement on the day when an outside storage yard is laid out. On that day, determine if the field modules will later become the permanently installed cabinets, warehouse spares, etc.
4. Negotiate for one of the experienced firm's experts to be the "quasi owner's" 50 hour-per-week on-site guide and consultant.
5. Sign a contract that explains that this project is a technology transfer and will involve your facility to receive copies of drawings and work procedures.
6. You may have to guarantee that you will not sell or otherwise transfer these documents to third parties.
7. Be sure the contract is clearly spelling out whether you do, or do not, have the right to be your own design and construction contractor if implementation of oil mist is contemplated in the future at your company's affiliates.
8. As a manager, periodically ascertain the effectiveness of your "quasi owner" in his/her role as the consultant's understudy for the duration of the project.
9. Ask your future "quasi owner" to write daily half-page status summaries, indicating parts consumption and percentage of project completion.
10. Have the "quasi owner" communicate these summaries to your organization.
11. Organize training sessions for maintenance personnel, followed by training sessions for operations personnel.
12. If more than one oil mist cabinet is involved, commence working on placing the second and third cabinets at their permanent installation sites.

13. Be sure to retrieve copies of work procedures and training materials from the guide/consultant. Catalog and save these materials and associated files in a computer folder.
14. Define and obtain spare parts. Assign a cataloging system that accepts/duplicates your company's existing identification and location-defining codes.
15. Realize that commissioning and handing over full ownership to your company must include all of the above. Do not allow delays in obtaining any of the agreed-upon deliverables.

Summarizing what we have learned: Pure oil mist represents the best-available lube application technology, and only jet-oil is in the same league. In terms of low maintenance cost and reliability, none of the traditional oil rings, flinger disks, and grease lubrication come even close.

Decide at the inception of a project if and how you will be your own implementation contractor. Arrange to have access to a „management sponsor." This sponsor should be called upon if disputes were to arise.

Part C: **Full equipment standstill/standby protection**

Chapter 8
Outdoor equipment storage and preservation yards

8.1 Management digest

The core setup for an equipment storage or protection yard is shown in Fig. 8.1. As many as eight, nominally quarter-inch-size tubing connections are provided on each of the 20–30 manifolds threaded into the common header. The 2″ header is slightly sloped (assume typically 0.5° to perhaps 1° off true horizontal), with the lower end of the header near the midsize OMG, in this case a small oil mist generator/oil reservoir combination unit. One of many manifolds is shown circled. This OMG can later be permanently installed in one of the facility's process units.

8.2 Overview and principles of storage yards

This chapter brings to the reader's attention that equipment staged in outdoor storage yards is serviced by the same oil mist generators (OMGs) that are used in permanent oil mist consoles or cabinets. The reclassifiers secured in equipment bearing housings are the same ones used once the equipment is installed in its permanent location. However, the owner-operator can request and specify that the outside storage OMG will later become a warehouse spare and that the very same warehouse spare OMG can be set up as an emergency module next to a machine that suffers from frequent bearing failures. Initially though, outdoor storage yards capitalize on the merits of oil mist as a preservation medium. We wish to point out that Appendix III also addresses the general topic. Appendix III is based on API RP-686, whereas Chapter 8 deals entirely with the author's decade-long experience.

8.3 Modifying new equipment upon arrival at a storage yard

Some minor sizing and configurational differences may exist between small and large storage facilities. Outdoor storage yards have been used throughout the world, and only a single US location ever used wooden rain and sun shelters for storage. The author strongly believes that storage shed decisions were made by lower level supervisors who, upon ultimate dismantling, transported the wooden beams and plywood panels to their small farms and weekend homes for re-use. However, an estimated 99% of temporary outdoor storage facilities were erected without such protective woodwork because such shelters are simply not needed.

https://doi.org/10.1515/9783110749441-008

Fig. 8.1: Core setup of an outdoor storage yard consisting of an oil mist generator module, 42-gallon drum of oil, header, and multiple manifolds (source: Don Ehlert).

As a general rule, equipment arriving in a storage yard for oil mist preservation receives the same preparatory treatment as equipment being readied for conventional, liquid oil-based preservation. Noteworthy facts are as follows:

- Oil mist will coalesce when the equipment operates. Operation causes a measure of turbulence, whereby the atomized particles get knocked together and become large and heavy. Turbulence is obviously not created in equipment that is standing still. Without turbulence, the reclassifying (conversion) of oil mist to liquid oil proceeds at a rather slow rate in non-running equipment.
- A low point drain location is being identified and left open at all times. This low point is useful because it prevents oil accumulation and ensures throughflow of a small amount of mist. The mist lost through a drain orifice equals the makeup mist.
- With rare exceptions, a 1 mm minimum, 3 mm maximum drain orifice is all that is needed.
- A machine in storage and ready for preservation can be slanted or skewed to create or locate a convenient and/or accessible low point drain. A flat pan can be used to catch this drainage for monitoring and proper disposal.
- A rotor blanketed by oil mist should be manually rotated two and a half turns every 6 months. This ensures a light coating of oil remains established throughout a rolling element bearing and avoids false brineling (indentation) at or near the 6 o'clock load points.

Only steam turbines require special cleanup before recommissioning. Accordingly, the following precautions apply only to steam turbines:

- Use oils with formulations that will not promote future stress corrosion cracking.
- Prior to machine commissioning, blow steam through the equipment. This removes the thin coating of oil that would have accumulated on the surfaces.
- Collect some of the exiting blow-down steam and observe if it contains oil.

If monitoring the discharge from a low-point drain is deemed too pedestrian, reasonably accurate calculation of residual oil films on machine interiors is feasible. As a not so general rule, one could proceed with an academic calculation. One could calculate the total interior surface areas that are being wetted in a machine and then assume that a 0.0005″ thick oil film will coat these surfaces. Knowing the area and oil film thickness, one could calculate the total volume of oil that will exist in the equipment in a time period and convert this to milliliters of oil, knowing that 3.85 L equates to a gallon. A liter equals 1,000 mL. Assume the premium preservative oil (Product A) costs $18 per gallon and figure out what it all costs after subtracting oil lost. One would have to account for oil lost through the 3 mm orifice, based on an equipment internal pressure of 0.1 to perhaps 0.3 psi above ambient and recalling that the oil-to-air ratio is 1:200,000.

Fig. 8.2: In this storage and preservation yard, crated machinery is initially accessed by removing one of the four sides of the crate (source: Don Ehlert).

A downward sloping plastic line (3–6 mm tubing diameter) is connected to each bearing housing and a drain plug (with a normally 1 mm and maximally 3 mm drilled hole) is installed at the bottom of the bearing housing. If the equipment arrives in crates as shown in Fig. 8.2, the crate must initially be opened and then reclosed after oil mist preservation commences.

Fig. 8.3: Cutaway image of a self-priming pump and electric motor driver, where all tubing connections are led to the two external panels (source: Don Ehlert).

However, opening the shipping crate can probably be avoided by specifying, and getting the manufacturer to provide, the external connecting panels seen in Fig. 8.3. Note, also, the care with which the pump inlet and outlet nozzles have been covered by this manufacturer. The crates illustrated in Figs. 8.2 and 8.3 are quite obviously ready for the oil mist tubing seen in many of the various illustrations that follow.

In case machinery arrives unprepared for outdoor storage, an EPC/owner's SME team will have seen to it that a small OMG/oil supply combination is standing by to protect the machines, Fig. 8.4. A proactive EPC/SME team will have ascertained that the principal OMG components can later become spares for permanently installed new or pre-owned oil mist modules. Components shown in Fig. 8.4 are ideal emergency lubrication modules for the occasional pumps that are the victims of lube design application errors.

Both Figs. 8.2 and 8.5 depict crated equipment that was purchased overseas and arrived after crossing an ocean. The captions provide additional information. Note the downward sloping tubing in Fig. 8.5; it prevents the formation of low pockets where coalesced oil could accumulate and impede proper oil mist flow. We recall that 2.31 ft = about 28″ of water column (~32″ of oil column) equals 1 psi. If the oil mist pressure inside a protected machine or vessel is 0.1 psi, a U-shaped 2.8- or 3.2″ tall low point/pocket in the tubing may act as a plug, and no oil mist will reach the intended destination.

Fig. 8.4: Outdoor storage setup with oil mist elements that can later be used as part of a permanent in-plant oil mist console (source: T. F. Hudgins, Houston, TX).

The oil mist application details shown in most of our images were originally specified by the user-purchaser. When the equipment arrived at its destination, the storage yard preparations seen in Fig. 8.2 greatly facilitated rapid hookup. Large and small crates were provided with the connections illustrated in Figs. 8.3 and 8.5. An oil mist preservation yard with equipment and installation per Fig. 8.5 may cost no more than $30,000 and payback will probably be realized in as few as 6 weeks. Note the midsized oil mist generator, new at about $16,000, a 55-gallon oil storage drum, and numerous 8″ tubing lines leading from manifolds to external panels on the crates.

An instrument-grade dry air supply with a dew point of −40° will be needed. After 6 months of this type of storage preservation, there are usually zero infant mortality events. Infant mortality typically refers to failure within 30 days of commissioning. Large machines are preserved with conventional products (more about this later) on the outside surfaces.

Vertical storage is illustrated in Fig. 8.5. Oil mist application is neither affected nor limited by the orientation of storage crates or the machines stored inside the crates. While the inside of a machine is fully preserved and protected with oil mist (see Fig. 8.6), the owner or field contractor may opt to apply further external weather protection in the form of plastic foil materials.

Fig. 8.5: Oil mist preservation applied on tall, crated equipment (source: Don Ehlert).

Recall, however, that oil mist protects bearings and other *internal* machine parts. Accordingly, after the application of a suitable external, grease-like product to *external* weather-exposed shafts, plastic sheets are customarily used to protect shafts and other external surfaces from the elements. The edges of the plastic sheets are secured with duct tape.

8.4 Preservation statistics and cost data

One of the two principal manufacturers of major oil mist systems provided important cost and size-related cost data.

To begin with, one of the two manufacturers has an operating manual comprising 25 pages, whereas the other manufacturer publishes a 100 +-page manual for a virtually identical unit. One of the two manufacturers confirmed that they produce only stainless-steel consoles and cabinets, whereas the second manufacturer offers both aluminum and stainless steel.

A 9-gallon console is offered for $42,000 without mist density monitor; $45,500 would include mist density monitoring. Back-mounted spares ore included with both offers. A bare-bones unit without mist monitoring or spare is only $34,500.

The vendor-manufacturer confirmed that $14,500 will buy a 5-gallon capacity unit with stainless-steel stand; $20,000 would be the outlay for a stainless steel cabinet on the same size unit.

Plain 3-gallon backup units sell for $3,900 but heaters are add-ons ($1,450) and a mounting panel costs $350. Finally, there are fully equipped units ready for indoor preservation duty at $17,500.

Past history listed an oil mist unit and associated piping that ably served the storage preservation requirements for ten gearboxes. It cost in the vicinity of $20,000; pricing was given as a budgetary cost estimate that included full installation. The installation was to use manifolds (see Fig. 8.1) and small tubing to reach from a header to the individual points to be protected (see Figs. 8.2–8.6). Smaller OMG units (see earlier, Fig. 6.5) can cost as little as $6,000. Uncrated equipment needs external protection, similar to that shown in Fig. 8.6. Internal parts are preserved with oil mist, and a grease-like product is applied to coat the external parts of the machine. Plastic sheets and duct tape prevent the coating from getting washed off by heavy rains.

Fig. 8.6: Uncrated equipment needs external protection, so oil mist coats the machine's external parts to preserve its internal parts (source: Don Ehlert, 2012).

Fig. 8.7: New midsize stainless steel oil mist cabinet (left side), and the vacuum-return oil tank with blower (right) (source: LSC, Houston, TX).

The estimate for the midsize OMG unit shown in Fig. 8.7 was about $16,000 when initially purchased from the original equipment manufacturer (OEM). It is part of a closed, unit-wide oil mist system, with the cost including the cabinet (also called a console), and its associated instrumentation. It does not include the return oil tank. Keep in mind that users involved in plant expansion or unit modernization often make smaller OMGs or OMGs with somewhat outdated or with less-than-perfect instrumentation available to used equipment dealers and buyers. Used equipment can be purchased and reconditioned at about 50–70% of new cost. It is frequently acquired to provide protection to major equipment stored indoors (Fig. 8.8).

Major suppliers of oil mist systems can furnish monetary data on the estimated overall economic performance for various plants. These suppliers can provide the data expressed as discounted cash flow (DCF) return, and also as payback period. Some suppliers divide the information into broad categories of user plants, including oil refineries, petrochemicals and polymers, ore mining, or perhaps, metals processing. Knowledgeable vendors can also provide technology updates, with details on the additional benefits of the technology.

For instance, as previously noted, there are significant benefits derived from using oil mist for both indoor and outdoor storage protection, as well as the mothballing (occasionally called hibernating and/or storage protection and preservation) of entire plants.

Fig. 8.8: Spare 9,000 HP drive motor for ethylene compressor; indoor warehouse storage with oil mist (source: Bloch-Ehlert, ~ 1982).

The large, 9,000 HP electric motor in Fig. 8.8 is a spare driver for a process gas compressor in Texas. Here, the motor is protected in an indoor warehouse, which the refinery calls stores or the primary location where its many spare parts are kept. No effort is made to confine oil mist to the two bearings only. Oil mist travels to all spaces and blankets or sweeps the entire interior of this motor. No moist air can intrude so long as the oil mist fills the motor at a pressure of 0.1–0.3 psi higher than ambient.

And yes, not all outdoor storage yards are giving a messy, albeit fully functional and entirely purpose-focused appearance. The neatly arranged equipment storage yard in Fig. 8.9 attests to the fact that a cost-effective outdoor installation can be all of the above plus accessible, space-saving, and neat in appearance.

8.5 Preview of alternative outdoor storage protection methods

In Fig. 8.10 the same oil mist blanketing (i.e., throughflow) approach is used on the inside of the lube oil reservoir, which is part of the compressor lube oil skid shown here and forms the transition to the next chapter.

Filling the reservoir with a premium-grade inhibited turbine oil that is later used for circulation and commissioning is feasible. However, the head space must still be filled with oil mist to prevent rusticles. Rusticles are stalactite-like growths formed when wrought iron oxidizes on sunken ships deep underwater. Therefore, oil mist

Fig. 8.9: Neatly arranged outdoor equipment storage yard (source: T. F. Hudgins, Houston, TX.).

blanketing is probably the most reasonable, simple, and cost-effective storage preservation method in use today.

Actually, reliability-focused owner-purchasers have two logical and perhaps equally acceptable choices: (a) Blanketing/sweeping the reservoir with oil mist or (b) filling it with an inhibited turbine oil. However, the latter choice would only be economical if the same charge of corrosion inhibited turbine oil is used during run-in, full speed trials and first-year operation of the compressor which it serves.

The air space at the very top of an oil reservoir for turbomachinery must still be filled with oil mist. Forgetting to fill the headspace can be very costly. Moist air may intrude and create the aforementioned "rusticles," droplets of rust-laden water that look like small icicles. The effects of improper or negligent preservation of head spaces in all types of equipment can have an adverse, long-term impact on a company's profitability. This point is important to remember because it is very rarely found in the literature.

Our coverage of storage yards, the many illustrations showing equipment in these yards, and even the projected costs, were based on oil mist as the filler or blanketing medium. Other media are feasible but are typically considerably more expensive, or more of an asphyxiation hazard if the space so blanketed is entered. Although best-in-class companies will prefer oil mist, other storage protection methods are feasible and will be highlighted next in Chapter 9.

Fig. 8.10: Compressor lube oil skid in outdoor storage; the reservoir is blanketed with oil mist (source: Don Ehlert).

What we have learned is simply this: Pure oil mist almost always represents best-in-class technology; only nitrogen blanketing may (occasionally) come close to qualifying as technically sound and cost-effective.

However, nitrogen blanketing may be expensive and, if filling large enclosed spaces, may become a deadly asphyxiation hazard.

Chapter 9
Storage protection use often followed by permanent installation

9.1 Management digest

The basis for storage protection by introducing an inert gas into equipment is easy to visualize. Filling a space with gas at a pressure higher than the surrounding atmosphere prevents the ingress of the latter with its generally present airborne dust and water vapor. The gas so introduced must be clean and the equipment casing and its bearing housing(s) may or may not be partially filled with oil. Economics play a role and the uninterrupted supply of an inert gas such as nitrogen must be ascertained. Safety being of paramount importance, users must make sure that people cannot possibly enter a nitrogen-filled space.

9.1.1 Important dual purpose of oil mist equipment

It is well worth remembering that modern plants include plant-wide oil mist lubrication but start out with an oil mist storage yard. Their setup will initially serve the temporarily stored equipment and the oil mist generating module will later be moved to serve an entire process unit. After installation in the process unit, it has occasionally served as many as one hundred or more equipment bearings. In another example, an oil mist storage setup initially serving 200 stored machines was later installed in a process unit where it supplied oil mist to only about 25 or 30 machines. In other words, once the stored machines are permanently installed or mounted on their respective foundations, the oil mist cabinet, which included a 10-L oil tank, an air–oil mixing nozzle and a supervisory instrument panel, was moved into a particular process unit for permanent installation and the controls adjusted for the final location. It would then serve to lubricate the bearings in, on average, 26 or more pumps and electric motor drivers within a radius of up to 180 m or about 600 ft. In some of these example cases, only the mixing nozzle in the oil mist cabinet had to be exchanged for a different size.

Engineers or technicians entrusted with the design of oil mist systems are usually tasked with selecting from a number of standard components. They seek sizing guidance from the owner-operator but have considerable flexibility of choice. The exact implementation steps taken to preserve or inhibit corrosion on inactive process machines depend on the type of equipment, expected length of inactivity, and the amount of time required to restore the equipment to service. On standby equipment (i.e., installed spares), oil mist serves as a protective blanket that prevents the intrusion of dust and water vapor for virtually unlimited periods of time.

https://doi.org/10.1515/9783110749441-009

9.2 N2 blanketing and/or nitrogen sweeping

Keeping in mind the preceding paragraph, covering or enveloping equipment with nitrogen is possible, but rarely considered economical in a process plant. It can be dangerous if individuals enter spaces without first fully venting the nitrogen. There have been cases of personnel suffocating when safety was overlooked. Nitrogen blanketing is generally applied in conjunction with the combination shipping/storing containers used for jet engines and similar equipment.

Long-term storage preservation by nitrogen purging is well known in the industry. Generally, this method of excluding moisture is used for small components, such as hydraulic governors or large components, such as turbomachinery rotors kept in metal containers. Nitrogen consumption is governed by the rate of outward leakage of this inert gas and may be kept at a low, highly economical rate if the container is tightly sealed. The container needs to be pressurized to only about 10″ (2.5 kPa) of water column, although a more typical rate is approximately 1–5 psi (7–35 kPa). Containers may be fitted with a safety relief valve to prevent over- pressuring. Alternatively, the container could be furnished with an orifice vent to promote throughflow of nitrogen at very low pressure. This is sometimes called nitrogen *sweep* with nitrogen flowing, or *nitrogen blanketing* if the gas is confined (pressurized) but not flowing.

Although there is this distinction, most people seem to use the two terms interchangeably.

While workers have suffocated in flowing and also in non-flowing nitrogen, nobody has ever suffocated in oil mist applied to stored equipment. Because safety is an ethical imperative, it must be included as part of the planning process. True reliability professionals will never allow safety to be brushed aside. That said, let's move on to oil mist sweeping, also known as oil mist blanketing.

Old habits persist, and so, as of 2021, there are still a few major machinery manufacturers that are instructing purchasers, owners, and customers to use nitrogen for storage protection. To be clear, the author's experience-based position differs. In 1975 and while on long-term assignment at a large construction site in close vicinity to Houston/Texas, oil mist was used for outdoor storage preservation, see Figures 8.1–8.6, also 8.9 and 8.10. Oil mist reached every nook and cranny of every machine that arrived at the site. Questions relating to steam turbines and exhaust condensers were addressed; the safety of personnel and full asset protection were of greatest importance. It was shown that what little oil coalesced inside a machine casing or vessel would drain by gravity action to the low-point drain plug and through its 1–3 mm (0.040–0.120″) hole.

It was reasoned (correctly!) that a "residual oil film" coating interior walls would (a) have a thickness of perhaps 5 micron (micro-meters), and (b) would be purged from the machine, steam turbine condenser, or other protected vessel during pre-commissioning. Among the typical pre-commissioning activities is the large-volume introduction of high temperature and velocity blow-down steam. Although only widely publicized after

1978, oil mist blanketing and/or oil mist sweeping has been done since perhaps 1960. It represents a widely accepted routine that is demonstrably less of an asphyxiation risk. Moreover, oil mist blanketing is a less expensive but flawlessly effective means of storage preservation in the open and also indoors. Whether indoors or outdoors, closed oil mist systems can be used to aspirate and collect oil mist exiting from equipment drain plugs. The coalesced oil is generally ready for re-use in oil mist systems.

9.3 Oil mist blanketing and/or oil mist sweeping

As previously noted, the most beneficial storage preservation method from a technical standpoint is immersion in oil mist. This is known as blanketing. An equally successful alternative is sweeping, which creates a constant throughflow with oil mist. Although this method has been used since the 1960s for equipment storage either indoors or outdoors, neither indoor storage in a building nor outdoor storage under a tent-like structure is needed on machines, valves and other equipment at standstill. Oil mist, if applied in a closed system, can be coalesced, gathered, and reused. The volume of oil and clean, instrument-quality dry air for preservation at standstill is roughly 10% or less of the oil mist that would be later required while the equipment operates. A storage preservation yard with large machines blanketed by oil mist at a pressure slightly above atmospheric (0.1–0.3 psi higher than the surrounding air) is shown in Fig. 9.1.

Fig. 9.1: Large machines with oil mist blanketing applied to all interior spaces and voids (source: Don Ehlert).

However, preservation and protection with a corrosion-inhibited turbine oil is also a contender if the oil used for preservation can be left in place. In other words, the same charge of oil would later be used in the operating machine where its function is, of course, to dissipate heat and to lower bearing friction.

Fig. 9.2: Oil mist sweeping/blanketing being applied to large and small electric motors in foreground (source: Don Ehlert).

Note that electric motors in outdoor storage preservation, shown in Fig. 9.2, and indoor storage/warehousing protection (shown previously in Fig. 8.8) also respond quite favorably to an oil mist sweep. There is no degradation of winding insulation in well-built motors that incorporate epoxy coated windings. No air can enter with oil mist, whether blanketed or swept.

9.4 Oil mist intrusion into electric motors

There is some history to oil mist and electric motors. Around the mid-1960s, most big-name electric motor manufacturers refused to submit proposals for oil mist–proof motors for fear of not knowing what will happen. The winning bidder was a small electric motor manufacturer in New Jersey. This bidder-manufacturer had purchased a rubber stamp with the inscription “Suitable for oil mist lubrication” and had so stamped every one of their standard drawings. They had ascertained in 1960 what took others 10 years to understand.

In the mid-1970s, and after having disqualified themselves from selling motors to some best-in-class user companies, a company known to the author finally caught on. They immersion-tested electric motor windings and observed the insulation at extreme temperatures. The tests corroborated what the New Jersey manufacturer had known all along: They proved that none of the lubricants used by process industries in their multitude of pumps and electric motors would cause epoxy-based motor winding insulations to degrade. As a result, an estimated 54,000 oil mist–lubricated motors are now (2021) in successful service. It should be noted that plant-wide oil mist systems can be cost justified even more easily if one’s assessment of the resulting cost savings includes electric motor lubrication.

What we have learned:

Storage protection use of an oil mist system is often followed by it being permanently installed in a process unit. Best-in-class petrochemical companies usually develop their engineering and project standards to take this dual use option into account. It represents best practice and greatest economy by far.

Reliability and cost optimizing user plants (best-in-class facilities) much prefer oil mist over grease in electric motor bearings. This preference exists regardless of whether the equipment is in outdoor or indoor storage. Of course, it fully applies to operating equipment and installed machines in standby mode.

A typical mothballing program for indefinite storage follows the practices outlined earlier and moves through the planned, budgeted, and executed stages. However, storage protection implemented after months of neglect or exposure to the vicissitudes of unpredictable environments makes no sense whatsoever. The next chapter tries to make that point.

Chapter 10
Why storage preservation as an afterthought will fail

10.1 Management digest

All the storage preservation recommendations in this book relate to mothballing (i.e., storage protection or "hibernation") of equipment that starts out clean. Chances are one gets away with implementing equipment preservation two or three weeks after the asset has been shut down or stopped. But there is a high probability that machines left unprotected for 6 months will not respond fully to any method of belated storage protection. If preservation finally gets under way after 6 or more months of zero protection, there will be a measurable shortening of remaining equipment life before major repairs become unavoidable.

10.2 When it is too late for storage preservation

If a machine is shut down and left unprotected for a few months before protective measures are implemented, it is usually too late. In wet climates, water vapors will migrate into machines and condense to become free water. Dust, dirt, and desert sand will likely intrude in dry climates. All too often, rust will form and/or sand will cling to previously lubricated precision parts, such as bearings. In these cases, precautionary dismantling and examination are the only reasonable courses of action. Flushing will generally improve situations involving contaminant ingress, but only dismantling, cleaning, and, in some cases, replacing parts will be an assured solution.

Once a machine is completely disassembled, its individual parts can be cleaned and reused if solids wash off easily. If there is oxidation and rust, rust remover gels are available, but using them will require decisions made with considerable discernment. The gels must first be carefully applied and later thoroughly removed. There is no general rule; each case must be decided on its own merits, whatever the cost. For every dollar "saved" by leaving machines unprotected for 2 or 3 years, someone will now have to spend 20 or more dollars. These are the likely expense multipliers even before the cost of start-up delays and the value of lost opportunities for profitable production and sales have been added.

10.2.1 How degradation progresses

Fluid machinery degradation is assumed to progress as shown in an empirical plot in the next chapter (see Fig. 11.1). Nevertheless, it is intuitively evident that leaving machinery completely open to the elements (both inside and outside of a machine)

https://doi.org/10.1515/9783110749441-010

will come at a price. One would not expect rusty machines to perform well in industrial processes after 3 years of unprotected storage. Therefore, the storage recommendations using oil mist, or the traditional methods described elsewhere in this book, do not apply to machines that have been left unprotected for over 6 months or perhaps even years.

10.3 The flushing option

Restarting machines after 3 years of unprotected storage is never recommended and may become a serious hazard to life and limb. The recommendations for equipment left unprotected in environments exposed to wet or fine solids for 6–12 months are relatively simple:

1. Flush the equipment with kerosene and drain the kerosene into a container.
2. Locate a linen patch (e.g., bedsheet-quality fabric made from fibers of the flax plant) on a 100-gauge wire mesh support grid between the equipment drain opening and the inlet to the kerosene catching container.
3. Examine the linen patch after every fill and drain cycle, looking for rust or dirt particles visible to the naked eye.
4. Continue and repeat the flush, drain and examine cycle until no more particles are visible to the naked eye.

There is no reasonable alternative to this flushing procedure. The speed of this procedure can be slightly accelerated by subjecting the machine's exterior to hammer blows. These blows will dislodge some rust that may have clung to ferrous components. But even then, the reliability professional must be prepared to find rust and/or corrosion pits which, in the case of rust, could signify internal parts damage. Corrosion damage cannot be reversed; precautionary dismantling and rebuilding efforts that consist of parts repair or replacement are recommended [37].

There are two other points to keep in mind:

1. In the unlikely event that a machine was designed and manufactured with O-rings that will swell or degrade when contacted by kerosene or commercial solvents, disassembly and O-ring replacement will be required.
2. A commercial solvent can be used instead of kerosene. Alternatives to kerosene and/or commercial solvents are usually offered by paint companies. Before accepting these commercial solvents or custom-designed products, reliability professionals should engage in an experience check with other users.

What we have learned:

Storage protection as an afterthought is always a colossal failure of management. Occasionally, the failure results from reliability professionals not making a sufficiently compelling case for the outsized benefits that can be derived from finding, allocating, or borrowing the relatively insignificant amount of money needed for an outdoor storage yard and associated provisions.

Chapter 11
CAPEX for best available technology

11.1 Management digest

All too often, the reward system for project managers encourages them to focus excessively on initial cost savings. Best-in-class corporations avoid upsetting the needed balance between keeping project costs in check when building the plant, while at the same time ascertaining its future safety, availability and profitability. These, of course, are among the key contributors to "runnability." Runnability is a collective term describing the built-in safety, quality, overall uptime capability, and long-term profitability. The need to build with runnability in mind is best conveyed to the project executive by the corporate higher-ups. They will inform the project executive that he or she will be involved in the plant's startup and thereafter will be in charge of operating and managing the plant for 3–5 years.

Promotions beyond managing the plant will quite obviously be a function of the plant's runnability achievements. It follows that storage preservation should be part of the capital budget and cannot be brushed aside by project executives. A project executive's budgeted asset cost should never be made on the basis of lowest possible cost of equipment. Again, runnability criteria must be invoked every step of the way.

11.2 Questions on funding

While maintenance funds are counted as operating expenses, the same is not the case for capital expenditures – capex, for short. The capex funds appropriated for new construction, occasionally called "grassroots" or "green field" projects, undergo different cost justification and approval processes. Moreover, new construction and the great majority of plant expansion projects receive different tax treatment. As a general rule, the upgrading of an existing asset undergoing maintenance or repair is legitimately buried in the operating expenses of a plant, the process unit or significant asset. Once upgraded, an asset will work more reliably or require less maintenance than it did before.

A similar rule can be envisioned as favoring the specification, procurement, and use of best available technology. This should be kept in mind whenever selection criteria place emphasis on assets with reliability-optimized future in-plant use in mind. It means that we may have to change our way of conducting business by adopting and emotionally supporting a better reward system.

https://doi.org/10.1515/9783110749441-011

11.3 Costs for small outdoor storage yard using a pre-owned OMG

Based on substantial amounts of information available since 1962, a strong case can be made for allocating capital for equipment storage preservation. Oil mist will be involved and the benefits far exceed the cost. As previously noted, the combined cost of a pre-owned oil mist generator (OMG), oil holding tank, air dryer and 10 hp compressor can be as low as $20,000. The price is based on black iron piping; galvanized or stainless-steel pipes are not required in oil mist systems.

Plastic pipe headers are unsuitable due to excessive sagging or temperature-induced movement. Not included are small-bore plastic tubing and connectors ($2,000). One hundred man-hours will be needed to do the installation and two experienced plumbers are needed for 1 week during the installation. At $80 per hour, the labor estimate is $8,000. Thus, the total amount of money needed (in 2021) for an outdoor storage yard is estimated to be approximately $30,000.

The probable cost of oil and air in the oil mist preservation case is thought to be completely balanced by the cost of oil and labor required for conventional storage protection. Therefore, it is not usually considered in the projected cost comparison or cost justification calculation.

However, the above calculation should be carried out only by personnel with prior exposure to efficiently executing the requisite work processes and procedures. Because the calculation assumes in-house professionals will have a working knowledge or will have taught themselves about oil mist, the cost of hiring a contract supervisor from an experienced oil mist provider is not included in the $30,000 estimate.

To determine additional savings, we often recommended developing a sample case (a cost workup) on paper. It would be helpful to contrast oil mist storage protection against leaving equipment unprotected for 1 year. Here is a conservative estimate based on one such case:

A hundred assets are involved. First, it was estimated that there would be two versus twenty-two failure events, which equals twenty avoided infant mortality failures by using oil mist storage preservation. At a projected repair outlay of $15,000 per infant mortality event, a plant will pay an incremental $300,000 dollars for equipment left unprotected for 1 year. Figure 11.1 can help in estimating probable infant mortality as a function of storage protection method and time.

The payback for oil mist being applied within two months of shutdown versus leaving equipment unprotected for 1 year is then estimated:

$$\$30{,}000/\$300{,}0000 = 0.1 \text{ year or } 5\text{–}6 \text{ weeks.}$$

Left unprotected for 3 years, all 100 assets in this example will have to be dismantled, cleaned, and portions of their components or parts replaced. In that case, the facility will likely incur $2 \times 3 = 6$ versus $100 = 94$ (incremental) repair events. At a projected

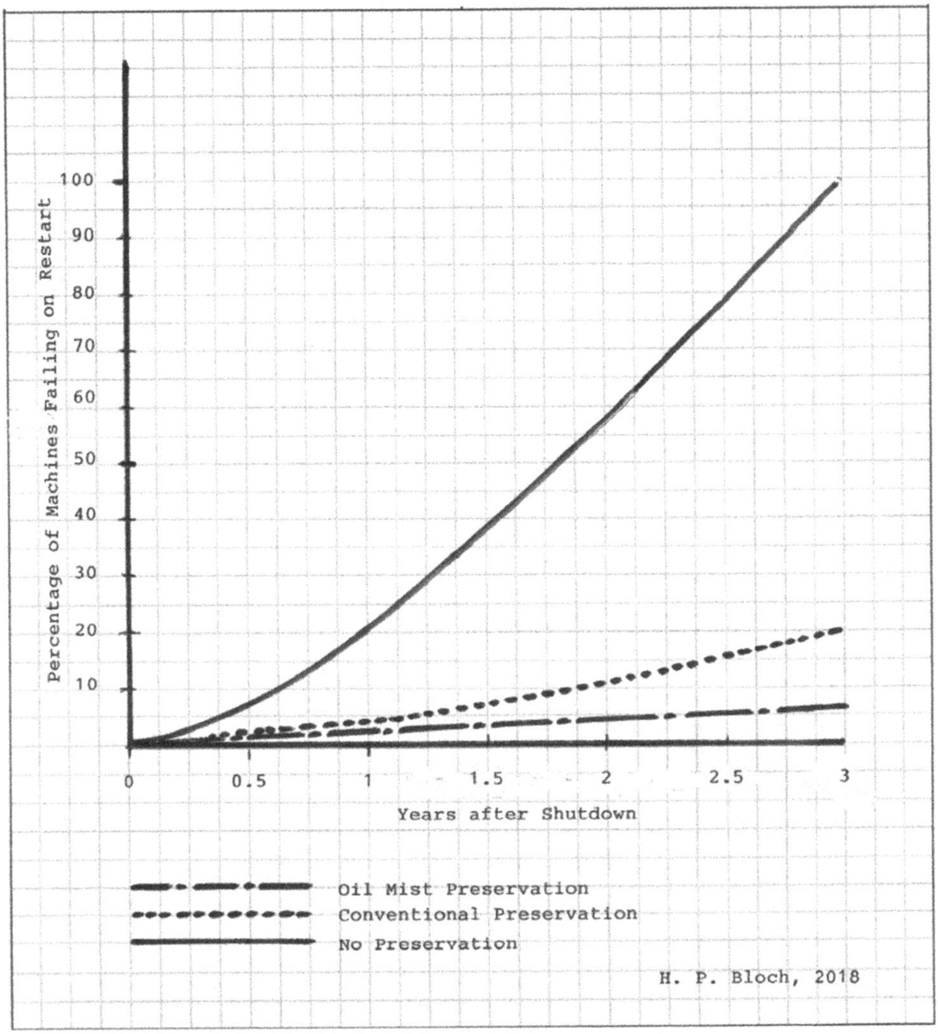

Fig. 11.1: Percentage of machines failing within 2–4 weeks of restart after shutdowns in wet or desert climates (source: Author).

dismantling and repair outlay of $15,000 per event, a plant will incrementally pay $1,410,000 for equipment left unprotected for 3 years.

The payback time for oil mist applied within 2 months of shutdown versus leaving equipment unprotected for 3 years is

$$\$90{,}000\,/\$1{,}410{,}000 = 0.064$$

This would be the fraction of 3 × 365 days = 1,095 days. Payback is seventy days, a little over two months. You can make your own projections by substituting or plugging in appropriate numbers.

However, we wish to stress the fact that the example we used here is based on an actual experience. We had been asked to advise a major US oil refinery that had received and then stored equipment in the open for 3 years. There was a huge storage yard, but there was no storage preservation. When the economy took a downturn, this refinery's internationally known corporate management had told local subordinates to immediately stop all work on a US $1 billion (870 million euro) refinery expansion. Roughly 3 years later, Corporate Managers asked the locals to go ahead and pick up where they had left off.

We emphasized the prudent course of proactively dismantling, cleaning, occasional parts replacing and reassembling many of the previously unprotected machines. We explained to the managers and senior staffers of this world scale refinery that they were now confronted with a series of unpleasant choices.

11.4 Costs for future large outdoor storage yards with factory-new OMGs

Using cost information in this book, it is quite easy to make the case for capital budgets that include proper outdoor equipment storage preservation. For this example assume a first-class, permanent, unit-wide system where it will provide state-of-the-art lubrication to seventy motor-driven pump sets. While half of these are on standby duty, they are fully preserved by oil mist. With the oil mist lubrication system initially doing outdoor storage protection duty, an estimated 300 assets will be preserved under an oil mist sweep that completely fills all casings and bearing housings in the storage yard.

In this case, the installed cost of a new oil mist generator and a new 500 gallon oil holding tank is estimated at $150,000. Arrangements were to be made for one of the unit's future instrument air compressors and dryers to be brought in early to provide the motive air for the oil mist supply in the storage yard. Scaling up and using the small yard/used equipment case as the pricing model, a total incremental cost of $200,000 is budgeted. Anything beyond that amount would be part of the plant's normal piping design and construction budget.

Again, the cost outlay is calculated based on the oil mist system being used initially in a storage yard and later transferring the OMG and its associated hardware to one of the facility's process units. There are firm plans to ultimately use it as a unit-wide oil mist supply source. Therefore, an all-inclusive cost justification would include using oil mist for pumps and electric motor drivers. A rigorous cost justification would take into account labor savings, considerable failure avoidance, and having fewer fire incidents.

11.5 Budgeting oil mist preservation

In early 2008, a Texas-based reliability engineer was surprised that the $4 million cost proposal he had just received for a plant-wide oil mist system was not close to the cost projected in an oil mist lubrication handbook he had read in 1987. Well, not only had things changed in the intervening 21 years, but it was clear from the ensuing conversation that serious misunderstandings tend to creep in when people do spotty and selective reading. The situation is even worse when project advisors relay their misunderstood tidbits of information as current fact.

The recommendations and actions of reliability professionals must be based on facts and these facts must be presented with integrity and without misleading, in any way, the reader, technician, or manager. Diverging views must be explained and reconciled. There can only be one fact, although it is possible to support an agenda and one's views or opinions by leaving out portions of the complete story. However, deliberately leaving out facts is dishonest; the individual who leaves off facts is a professional in name only. Of course, comments by an uninformed, careless, or indifferent person may inadvertently leave out facts. But, irrespective of whether purposeful or inadvertent, omissions can be avoided, and true professionals will not become a party to the dissemination of misinformation.

Which leads us to the topic of context. Diverging views can, for the most part, be reconciled by looking at full context. Under the next subheading, we will briefly examine why context is important.

11.6 Why context matters

In 1990, the presenter at a technical conference showed images of a large number of different process pumps with stainless steel (hydraulic) tubing connected to each bearing housing. Other images depicted outdoor storage yards with plastic oil mist tubing. It was noted that plastic tubing is perfectly acceptable for stored equipment, but stainless steel tubing was needed in a process unit. The reason why stainless steel tubing was to be specified for machines within the boundaries of a refinery process unit had to do with fire issues. Heat from fire in a nearby pump could melt plastic tubing in an otherwise unaffected machine. The real message tends to be lost or badly misinterpreted if all that we remember is the partial sentence "... plastic tubing is perfectly acceptable."

Twenty years earlier, in the late 1970s, an American Society of Mechanical Engineers panel discussion arranged for four panelists to report on the successes of plant-wide oil mist lubrication. One of the panelists worked for an offshore employer; he spoke about issues with oil mist at his plant location as he projected several 35 mm slides on a large screen. The other panelist immediately pointed out that the plastic (!) tubing runs from the various manifolds to their respective bearing housings were too long. They sagged and allowed low spots to exist. Coalescing oil collected in these low spots and prevented oil mist from reaching bearings. That refinery's newsletter reported that oil mist systems are prone to experience unexplained random bearing failures but omitted spelling out the reasons, which were in plain sight.

There is not enough time here to relate even a fraction of the many tales and anecdotes that are being passed along over a few beers at barbecues. Examining facts and becoming familiar with the underlying science makes eminent sense. Facts are worthy of being brought to the attention of every functional layer in a reliability-focused organization. Opinions, anecdotal banter, and sentences quoted out of context can pose serious risks to the safety and profitability of a plant. When opinions take over, deviations from the original quality norm tend to become the "new normal" and repeat failures are experienced.

11.7 Thorough cost justifications require study of statistical information

Figures 11.2–11.10 will be helpful in cost-justifying a plant-wide system that includes 600 pumps and a storage yard. The before and after oil mist costs were compiled with input from a user's consultant, an oil mist provider, and from discussion group attendees at two Texas A&M International Pump Users Symposia. The figures include experience-based data that predict and/or give due credit to anticipated maintenance cost avoidance and the savings that accrue from failure avoidance. These cost comparisons can be adapted to local circumstances and/or using data from one's own plant. They can also be scaled up or prorated using reasonable and applicable ratios.

Considerations involving pump bearings (Fig. 11.2) and mechanical seals (Fig. 11.3) are rather obvious. However, the substantial savings through oil mist lubrication of electric motors (Fig. 11.4) are often overlooked.

Motors have weep holes at one of the lowermost points in the frame. Oil mist and coalesced oil can escape from there without causing winding insulation to degrade. Since the mid-1960s, irradiation cross-linking polymer insulating tape was chosen for the T-leads of electric motors because of its low swelling rate. Pre-1960s motors occasionally, but very rarely, have been found with common electrical insulating tape exhibiting high rates of swelling. In those cases, the pathways for cooling air may become blocked and motors will overheat. This is simply an example of why it is important to work with a motor manufacturer who understands why good motors use the state-of-the-art insulating tape material. Do not pick the cheapest possible vendors for your electric motors. Understand motor lubrication, regardless of grease, liquid oil, oil mist, oil jets, and so forth.

A good cost justification study will not neglect manpower savings (Fig. 11.5), extremely high lubricant savings with closed-loop oil mist versus conventional lubrication (Fig. 11.6), and the value of fewer repairs translated into fewer fire incidents (Fig. 11.7).

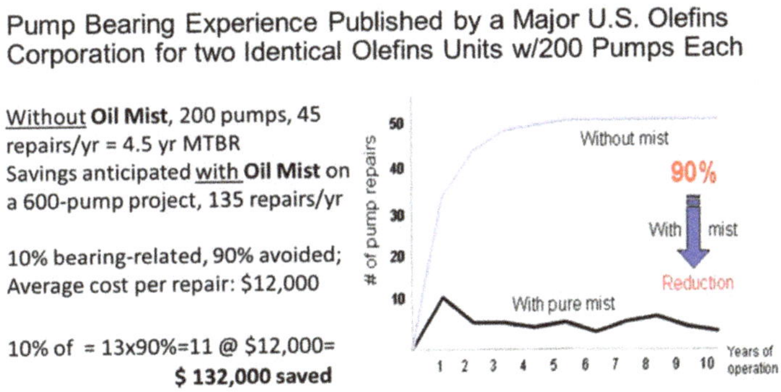

Fig. 11.2: A prominent user collected these data from two identical process units, one used oil mist lubrication, the other the vendor's standard lube methods (source: Charles Towne/Shell & Don Ehlert,/LSC Houston, TX).

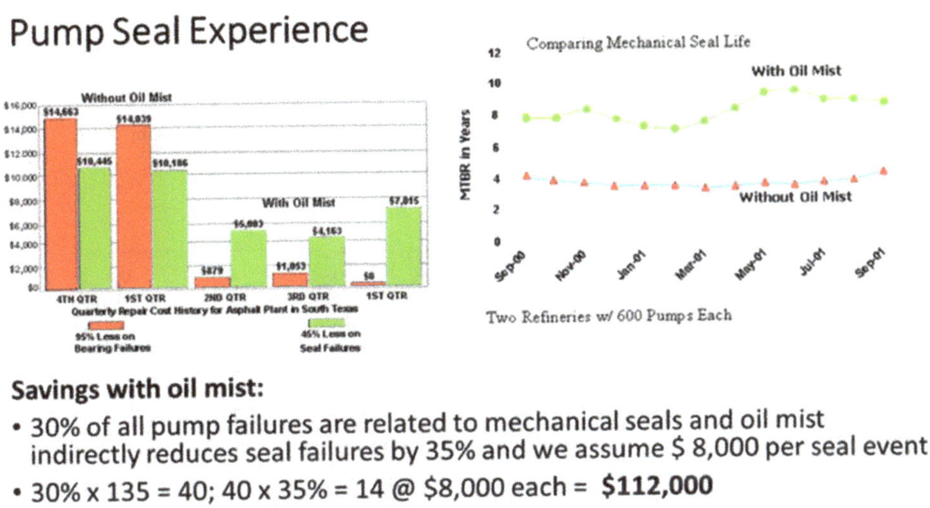

Fig. 11.3: Bearing distress affects mechanical seals, oil mist lubricated bearings indirectly contribute to seal failure reductions (source: H. P. Bloch and Don Ehlert).

11.8 Summary of findings and how data are validated

Many companies find the cost to invest in oil mist is justified, although their calculations are often based on lower pump maintenance costs alone. While in themselves noteworthy, savings in driver maintenance and repair (no more regreasing of motor bearings!), lower operating manpower, reduced lubricant consumption, energy savings of 2% (although these savings are considered offset by the cost of compressed air). Then there are potential production gains (savings, also known as opportunity profits, Fig. 11.8) that merit consideration as well.

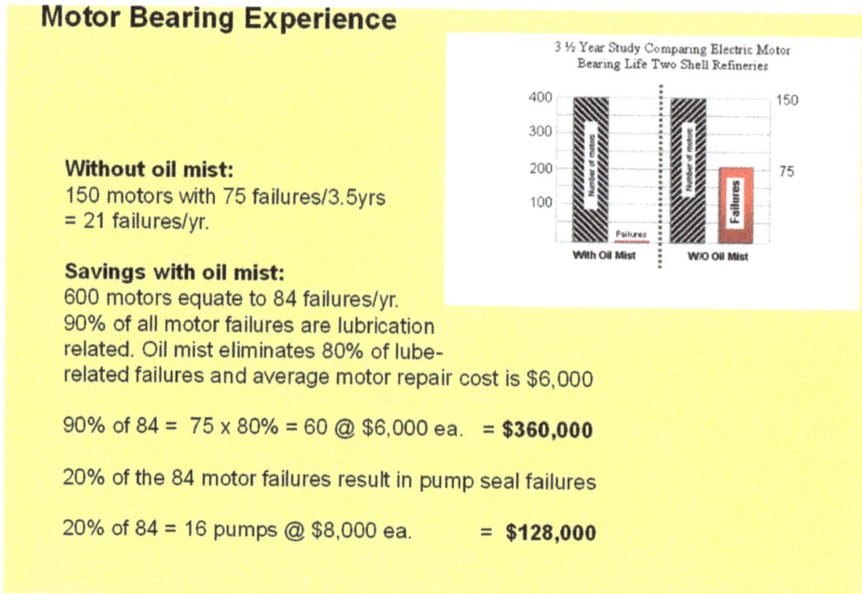

Fig. 11.4: Ninety percent of electric motor failures originate from distress in grease lubricated bearings (source: H. P. Bloch and Don Ehlert).

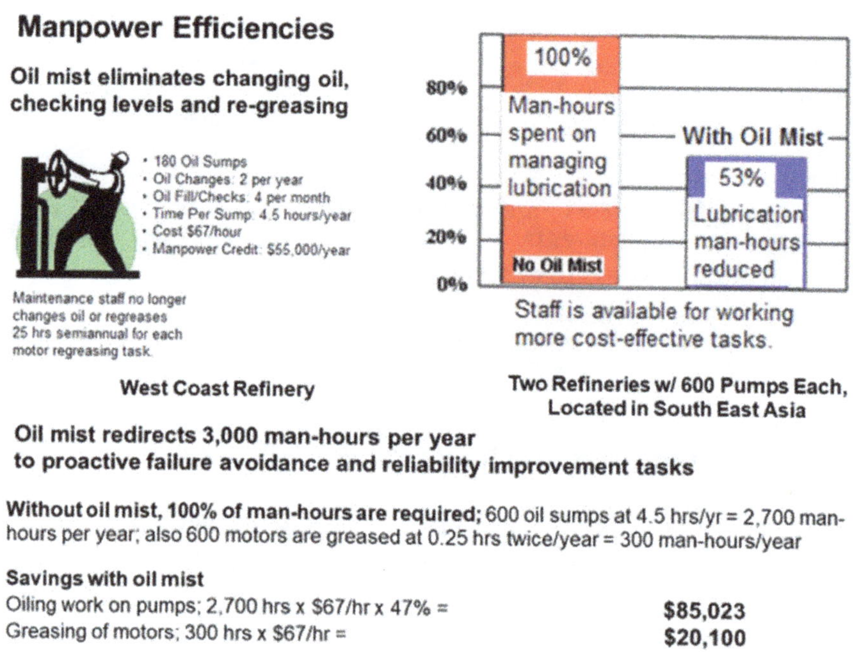

Fig. 11.5: Oil mist lubricated machines require only 10% of the maintenance needed for conventionally lubricated equipment, freeing up labor and production (source: H. P. Bloch and Don Ehlert).

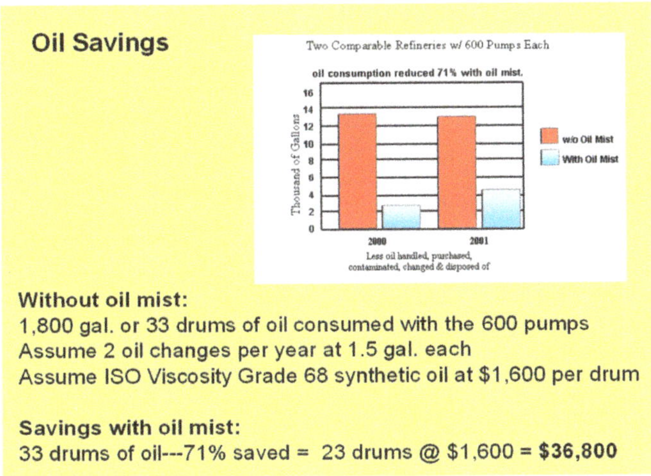

Fig. 11.6: Oil consumption is reduced with oil mist (source: H. P. Bloch and Don Ehlert).

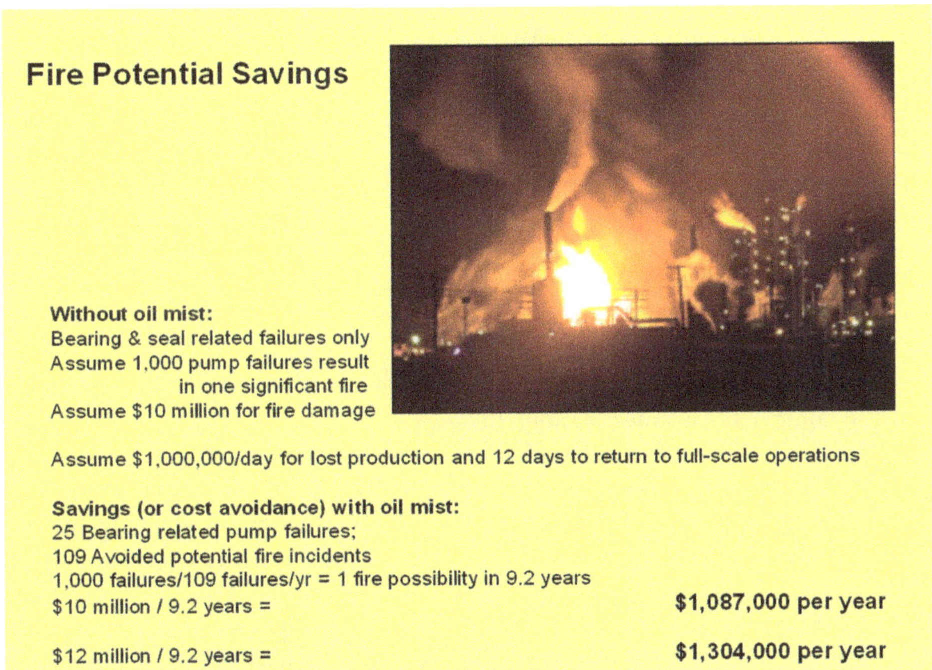

Fig.11.7: Fewer pump failure events mean fewer pump-related fires (source: H. P. Bloch and Don Ehlert).

Potential Production Savings, Non-Fire Related

Without oil mist:

Spared equipment failures of pumps, motors, turbines that result in minor upset of production
Assume $1,000,000/day loss for any given process unit
Assume 5 days lead time for repair; eliminating 109 failures = 545 additional days/yr for production
4.5 yr/MTBR leads to a failure every 1,606 days for any given service

Lost production avoided with oil mist:

545 days / 1,606 days = .33 additional production days/year
$1,000,000 x .33 days = **$330,000**

Fig. 11.8: Value of production loss avoidance where pumps have spares installed on an adjacent foundation (source: John Hartmann and Don Ehlert).

Oil refinery statistics show one fire per one thousand pump failures. The detailed statistics of major oil companies discussed during two Texas A&M University Pump Users Symposia and examined by the multinational machinery engineering network, NAMNET, in the two decades from 1965 until 1985 showed oil rings and incorrect oil levels in pump bearing housings to be chiefly responsible for bearing and/or mechanical seal failures that resulted in pump fires.

However, neither constant level lubricators nor oil rings nor sight glasses or desiccant breathers are used with pure oil mist lubrication. It is these components that are most often involved when pumps fail. Impeller failures are rare, in comparison. With fewer pump failures, there will be fewer fires. Once the value of avoided fires is included, the results significantly amplify the strongest case for oil mist lubrication. Production credits deserve to be included, as is the obvious value of reducing the frequency of fire incidents. Both must be factored into payback calculations. Production-related savings and losses are assessed in Fig. 11.9 and the entire analysis summarized in Fig. 11.10.

Unless proven otherwise, and there are such rare instances, plant-wide oil mist is cost justified for new plants and retrofits. A sound, strategic, technology package is aimed at maximizing reliability and minimizing lifecycle costs for rotating equipment. These lube-related reliability improvement efforts bring factual information to

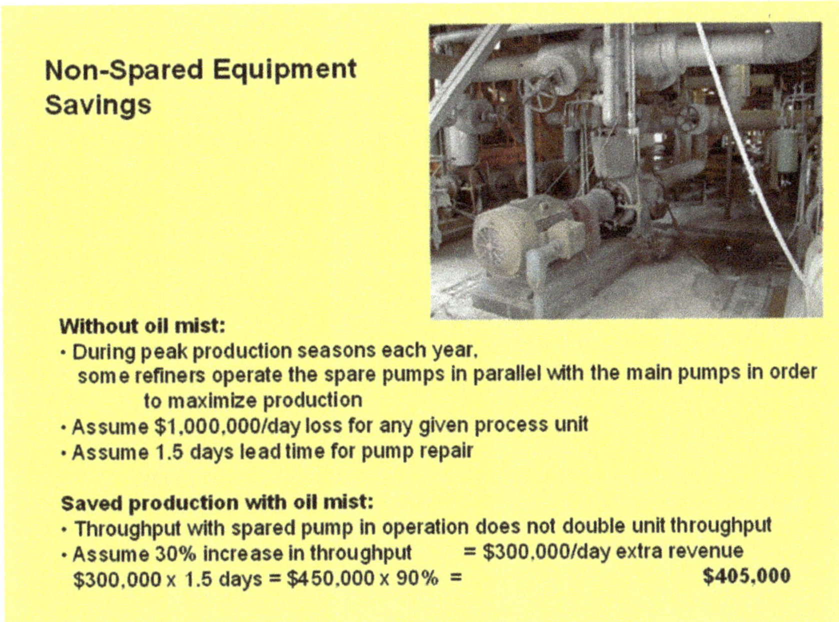

Fig. 11.9: With no spares, equipment failure will cause downtime and lost production (source: H. P. Bloch and Don Ehlert).

reliability engineers so they can reconfigure their data into a formal cost justification and bring it to the attention of their management. Formal cost justifications can be facilitated with the help of, and solid input from, knowledgeable vendors. Such an analysis might quickly prove the true value of a multimillion-dollar plant-wide oil mist project. We can think of situations where additional savings can be found and where the reader or analyst will see fit to add to or amend, or modify the cost bases highlighted in Figs. 11.2–11.10.

We can state without hesitation that reliability professionals who purposefully engage in relevant literature searches and then cost justify their recommendations add value to an enterprise. Reading and cost-justifying are activities that allow these professionals to more easily convey the merits of best available technology to their employers. Doing so will not detract from their principal daily role of keeping the plant running safely and reliably. It will, however, assist reliability professionals to become more productive, resourceful, informed, valued, and authoritative contributors.

As our concluding remark on the topic of CAPEX: Consider including a dedicated large tank to hold 2 years' worth of lubricant for your oil mist systems. The grainy photo of this tank (Fig. 11.11) dates back to the year 1978, which is when the plant was originally commissioned. The money for this large 6,000-gallon tank was in the budget.

Savings Summary

• Pump Bearings	$132,000
• Pump Seals	$112,000
• Motors	$488,000
• Manpower	$105,123
• Oil Savings	$36,800
• Fire Potential	$2,391,000
• Production	$330,000
• Non-Spared	$405,000
Total:	**$4 Million ($3,999,923)**

Fig. 11.10: Payback for oil mist lubrication is often within 12–16 months of implementing and commissioning a system (source: H.P. Bloch and Don Ehlert).

Fig. 11.11: A 6,000 gallon oil mist lube storage tank (source: H. P. Bloch).

The facility originally had 14 plant-wide oil mist systems of varying capacities, some 5, others 8 or 10 gallons. Whenever a cabinet-internal reservoir had used up most of its oil, a level sensor called for the tank's outlet valve to open. A small oil pump started automatically; oil flowed in the oil supply header and the branch line leading to a near-depleted cabinet-internal reservoir. Once 90% of full level was reached, the reservoir level sensor called for certain valves to close and the small oil pump near the outlet of the oil storage tank would be stopped.

After two major expansions the plant now produces almost three times the original yearly nameplate tonnage of olefins. Other oil mist systems were being added and older ones converted to closed systems. Less oil is needed today than in 1978 and the size of ZTK-27 is probably still just right. Its owners owe a debt of gratitude to the foresighted planning of the original EPC team and a senior manager with a clear vision. He made it his mission to look into the future and is fondly remembered by many. The exemplary facility he built was listed at or near the very top of best-in-class producers year after year until today, 43 years later, in 2021.

What we have learned:

Best practices involve familiarization with the steps and procedures that have allowed the competition to prosper. Best practices involve digging up facts, leaving aside anecdotes and tales of wishful thinking. Best practices require cost justification studies and acknowledging that everything we will ever need to know about optimized lubrication and storage preservation has been studied before, reported on before, and written about before. If the source is trustworthy, accept the advice.

With a near 100% probability that our assets are identical to assets elsewhere in our industry, there should be no need to ever engage in trial and error solutions. Conversely, there is a near 100% probability that reading will be required to discover what others have known, have done, or have published.

Chapter 12
Can field trials be bypassed?

12.1 Management digest

In 2014, a company in South America insisted on field trials before accepting their reliability engineer's recommendation to solve bearing problems with better lube application strategies. This chapter establishes that reinventing the wheel is not necessary in the face of facts that are readily available.

12.2 No field trials needed for oil mist

The success of oil mist lubrication on a scale ranging from just one machine to 30 plant-wide oil mist systems serving 2,000 or more machines at a single site has been a matter of record since the mid-1960s. Oil mist experts with access to many plants and reliability professionals with decades of applicable experience consider oil mist lubrication a fully proven and superbly reliable technology. They have become experts by learning from others, being inquisitive, and considering it their obligation to read books, articles, and conference proceedings on the subject matter.

Not one of these experts would find it plausible to request or endorse field trials. Even today, field trials would add nothing to the existing knowledge on the subject. We mention it because the issue comes up when uninformed persons call for field trials to prove that oil mist lubrication is not just wishful thinking.

On the two known occasions when field trials were requested in the 35 years from 1986 until 2021, there has been no justification for such demonstrations. Perhaps it was a subcontractor's attempt to turn experience-based technical advice into a "make-work" project. Or it could be that "seeing is believing" and the individuals making the "show me" request hoped they would not have to spend time reading. Whatever the case may be, it is inappropriate and wasteful to demand proof of concept and/or selection of beta sites for mature technologies such as oil mist lubrication and oil mist preservation. Hundreds of full-scale installations attest to its effectiveness and reliability. Mature technologies are the exact opposite of ideas or pursuits that warrant prototyping.

12.3 Field trials for conventional storage preservation

With regard to conventional methods of equipment preservation using products A, B, and C (covered in Chapter 13), the suitability of using these protection methods from a pure technical point of view is undisputed. Competent manufacturers and vendors

https://doi.org/10.1515/9783110749441-012

have continually improved these products. Their product summary books and applicable brochures are informative and helpful. All the available brochures are carefully reviewed and, where needed, updated. These updates consist of occasionally clarifying or amplifying some of the recommendations issued by lube marketers when they align with the field experience of experts.

Today, lubricant technology is accessible worldwide. Accordingly, the recommendations and summaries of major US providers of industrial lubrication could probably be joined by similar summaries available in electronic form from providers in other parts of the world.

12.4 Definition of deliverables

Expert providers of oil mist lubrication and other preservation technologies usually have access to transparent polycarbonate or plexiglass (acrylic) demonstration models. For example, a transparent bearing housing replica equipped with ball bearings and a steel shaft is an ideal visualization tool. Mechanics, operators, supervisors, and managers can readily observe oil mist in operation.

Generally, expert providers would agree to develop, sign and adhere to contract clauses that show machine interiors in the as received versus the as removed from storage condition. The provider would have a service contract with the client and would be able to monitor the storage yard, quality of instrument air supply, adequacy of lubricant, and so forth. The contract terms may include corrosion monitoring details and an up-front definition of remedies.

What we have learned:

Best practices do not involve re-inventing the wheel; they do, however, require familiarization with the steps and procedures that have allowed the competition to prosper.

Familiarization with best practices may be facilitated with demonstration videos, and/or scale models. These are often available at minimal cost outlay. Some can be purchased, others rented or leased.

Chapter 13
Vapor-related and old-style conventional storage protection methods

13.1 Management digest

A number of oil refiners and related manufacturers offer vapor phase and vapor space inhibitors to simply complete the range of corrosion-inhibiting products in their marketing portfolios. End users have a tendency to favor marketers who can offer anything of value to manufacturers, distributors or consumers seeking to inhibit corrosion or to prevent corrosion attack.

Old-style conventional storage protection methods still have their place but should be used only after understanding why their use is on the decline. These, too, are described in this chapter.

13.2 Examining vapor phase and vapor space inhibitors

Volatile corrosion inhibitors (VCI) are also known as vapor phase (VpCI) and/or vapor space (VSI) corrosion and rust inhibiting products. VpCIs volatilize at room temperature, forming a thin deposition layer on nearby surfaces. Some products can be sprayed into a space or volume; others come in powder form and are manually deposited in the space they are intended to protect.

Some vapor phase inhibitor (VPI) compounds are nonreactive and simply provide a thin barrier that minimizes oxidation. Others in the VPI family of products actually react with acids in the space or volume they are protecting.

Certain VCIs are hygroscopic, meaning they attract water. When thinking about using these products, it is necessary to have a clear understanding of the locations and configurations you wish to protect, how often the product must be replaced or replenished, and what the unknowns are in the equation. Perhaps this is why many reliability professionals have heard of VCIs, but have never actually seen them used in the petrochemical and refining industries.

While VCIs are produced and sold by reputable companies for the industrial market, the consulting advice given to potential users is quite simply: Trust, but verify. You should:

- Use VCI products only if you manage to prevent the vapors from escaping. This means you must ascertain that all escape paths are sealed and secured with tape.
- Consider if shaft systems are involved and the shafts must be turned periodically. If so, then breaking and resealing and re-taping with rust inhibited (impregnated) tape tends to become a big and often very labor-intensive maintenance job. Unless entrusted to dedicated workers, the results may prove unsatisfactory.

https://doi.org/10.1515/9783110749441-013

- Ask the VCI's producer and its associated vendors to put you in touch with other users that have been successful in using VCIs in the same types of machines you are trying to protect.
- Ask if the satisfied users' environmental conditions (e.g., humidity, temperature, airborne dust) were reasonably close to your conditions.
- Ask how often the product was reapplied in any particular time period. Ask questions about toxicity.
- Let the vendor explain removal from your assets prior to recommissioning after long periods of storage.
- Fully involve your resident technical staff in the decision-making process and obtain their buy-in.
- Meet with the employees who will be given the responsibility of ascertaining that all requisite action steps will be carried out faithfully. Don't allow someone to unilaterally use shortcuts. Realize that this meeting is of great importance if the decision affects another department or organizational jurisdiction.
- Consider what is necessary if you plan to use contract employees. Be sure to include them in training sessions organized, provided, and conducted by the VCI producer's sales force.
- Put the disclosure of experience and seller's training obligation in the specifications appended to your company's purchase order.

13.3 Opting for conventional storage preservation and selecting products

It may be difficult to envision physical asset location, dimensional and material-related makeup of a specific asset, usage details, and sets of circumstances where conventional "old style" storage preservation makes economic sense. We have last seen elementary short-term preservation steps taken on farm equipment in 1948, on liberty ships anchored in the Hudson River in 1953, and on military hardware in Arizona and Louisiana in 1955 and 1956. Nevertheless, a South American mining industry consultant inquired about quick and easy storage protection in 2018. Accordingly, the author decided to include conventional storage protection in this comprehensive text. All the major oil companies offer three families or product slates. These families are known as product A, product B, and product C. The consultant was asked to pick from these product families.

13.4 Properties of product A

Product A is a rust inhibitor circulation oil with an ISO viscosity grade (VG) of 32. In essence, product A is a turbine oil to which a special formulation or percentage of anti-foaming agents, oxidation inhibitors and three-phase rust inhibitors have been added. A machine requiring a turbine oil can probably be operated at its normal rating when using product A. However, contacting the equipment manufacturer would be recommended if product A substantially deviates from the manufacturer's stipulated oil type and performance parameters.

Product A and similar oils are made by major lubricant marketers whose respective formulations remain closely guarded secrets. The three-phase additive package is really (a) a rust inhibitor intended to protect surfaces that are lightly coated with oil, (b) a liquid phase inhibitor intended to protect surfaces submerged in oil, and (c) certain vapor phase inhibitors intended to protect surfaces that are exposed to oil vapors.

Oil companies are usually careful not to risk lawsuits. That is why one can be reasonably certain that the advertised claims of at least one major producer of storage protection fluids are correct. This producer states that product A, an enhanced or fortified turbine oil with ISO VG 32:
- Protects steel surfaces in the vapor (air) spaces that do not come into contact with the oil.
- Gives rust protection for metal surfaces from which the oil has been drained and where the product leaves a thin residual coating or film.
- Incorporates demulsibility properties similar to those of a turbine oil.
- Protects surfaces that are submerged in the oil.
- Protects other metals occasionally found in machinery.
- Has superior oxidation stability.
- Is suitable for protecting a wide range of equipment in a wide range of temperatures.

There is usually a caveat for product A: The vapor phase inhibitors will rapidly deplete at temperatures above 120 °F/49 °C.

Product A and its nearest competitors have the following principal properties:

– Specific gravity	0.88
– Pour point	20 °F (–7 °C)
– Flash point (ASTM D92)	380 °F (193 °C)
– Viscosity index	95
– TAN (ASTM D664)	0.4
– Demulsibility (ASTM D1401)	0
– Rust protection (ASTM D665)	Pass, distilled, and/or seawater

13.5 Properties of product B

Product B is an oil and is more viscous than product A. Operating certain machines with product B in the lubrication system or oil sump is possible for short periods of time, but oil rings may no longer qualify as the lube application method because of high viscous drag. Most oil rings are designed for ISO VG 32 and will not turn properly in the higher viscosity product B. However, product B provides effective rust preventive films on the internal surfaces of machinery, and it is this capability that makes it a highly suitable storage protection fluid.

Generally speaking, product B oils displace water from metal surfaces and form relatively strong, water-resistant films on metal surfaces. Corrosion and rust cannot form while the film is intact. If water is already in the system, product B absorbs the water, thereby turning it into a water in oil emulsion. Therefore, the surfaces remain protected.

In most applications, the residual oil does not need to be removed by flushing when recommissioning the machinery. There are exceptions to this rule if the:

- Manufacturer or vendor of product B finds it incompatible with the lubricant recommended by the equipment manufacturer. In this case, flushing with the recommended lubricant would be advised.
- Residual quantities of product B, regardless of the lubricant application method, would significantly alter the viscosity and temperature performance of the regular or recommended lubricant. An example is a high-pressure hydraulic system where water in oil emulsions could reduce anti-wear properties of the system's oil.
- Draining of the machine's casing or sump is difficult due to its internal geometry or construction features (pockets).
- Presence of the rust preventive agents would reduce the ability of the new oil to shed water. This is important because the lube oil also serves as the motive fluid in hydraulic stroking mechanisms and actuators.

Product B and its nearest competitors have, approximately, the following principal properties:

– Specific gravity	0.92
– Pour point	– 20 °C
– Flash point (ASTM D92)	~ 190 to 210 °C
– Viscosity index	95
– TAN (ASTM D664)	0.4
– Demulsibility (ASTM D1401)	0
– Rust protection (ASTM D665)	Pass, distilled, and/or seawater

Overall, product B are lubricants that also function as rust preventives. They are especially useful if machines are subjected to repeat cycles of operation under moderate load, then followed by a shutdown. Except for the aforementioned instances, the use of product B can save the costs of cleanup or flushing prior to again placing the equipment in service.

13.6 Properties of product C

Product C is a rust prevention, highly viscous, grease-like substance that is intended for applications with external components, such as exposed shafts, couplings, and wire ropes. It is a premium rust prevention grease that can be sprayed or brushed onto a variety of surfaces. Product C exhibits excellent water displacing properties and forms thin, tenacious films. These films protect surfaces, even under high moisture conditions. They are pliable at –35 °C and will not drip at temperatures up to +60 °C.

When formulated to provide wet, oily, and extremely thin barrier films that are not tacky, product C will not unduly attract dirt or dust. Product C also survives light acid fumes and serious exposure to salt-laden moisture environments, such as those encountered on a ship.

A slightly different formulation of product C forms a grease-like film that protects wire rope from corrosion. It stands to reason that this particular formulation should be sought out regardless of whether the wire rope is in operation or dormant, such as laid up and not moving for any length of time.

Product C and its nearest competitors have the following principal properties:

- Specific gravity 0.86–0.88
- Pour point –7 °C
- Flash point (ASTM D92) 120–240 °C
- Dropping point, ASTM D2265 up to 63 °C

What we have learned:

Best practice companies and/or professionals advocate vapor-related and old-style conventional storage protection methods for short-term preservation of assets only.

Grease-like conventional products can be obtained with viscosities that allow pouring the product into cavities or brushing it on external parts. In general, these products are applied while equipment is in transit from the manufacturer to the purchaser. Working with experienced product manufacturers will prove advantageous.

When trying to extend the period of protection to reliably last longer, vapor-related and old-style conventional storage protection methods will require potentially costly re-application and maintenance. Their removal and proper clean-up will incur added expense later. Once these facts are considered, it will often make vapor-related and old-style conventional storage protection methods unattractive for long-term preservation tasks.

The use of vapor-related and old-style conventional storage protection methods will be a risky proposition once risk-taking decision makers get involved. Suppose the reliability professional went along with these methods under the assumption that storage preservation was needed for 3 to at most 6 months only. Now the pro may be without recourse when the risk-taking managers claim there is no budget item for maintenance, product replenishment, shaft rotation, or re-caulking. Chances are that he/she will be blamed for the ultimate disappointments.

Chapter 14 Machine-specific storage preservation steps

14.1 Management digest

Needless to say, there will be substantial differences between storing a home sewing machine and preserving a surplus commercial airliner. We remind the reader that our comments and experience-based recommendations are confined to the machines found in modern industry. Of these machines, the number of electric motors far outweighs even pumps, the next largest category of rotating machines. Bulleted guidance is given here for strictly "old style" protection.

14.2 Small motors and similar machines

Usually, grease lubrication is found on small motors and the bearing style can be discerned from code letters on the motor's nameplate. However, not all motor manufacturers use the same letter code for their motor bearings. The Internet is a valuable source of information, as are the marketing departments of the different motor manufacturers.

When using traditional, often uneconomic means of preservation on small motors, it is best to keep the following 11 "bulleted" guidelines in mind:

- If the code letters indicate sealed bearings, they have been provided with lifetime lubrication and cannot be regreased.
- Unless it is possible to connect to an oil mist line, disregard the oil mist labels at all four locations in Fig. 14.1.
- At the two bearing housing lube outlet ports, connect pipes or tubes exactly as shown in Fig. 14.1. Unless connected to a closed oil mist system, these ports will always remain open, both while in storage and in actual operation.
- Thread grease fittings (also known as grease nipples or Zerk fittings) into each of the two bearing housing inlet ports.
- Use the exact type of electric motor (EM) grease that was previously supplied with the motor and apply it with a grease gun.
- Slowly apply grease to the two inlet Zerk fittings until the grease leaves from the horizontal pipe or tubing portion at each of the two outlet ports.
- When new grease arrives and falls out, discontinue the grease application.
- With heavy-duty pliers, squeeze the two open outlet ports into oval shapes with minor diameter openings about 0.100″; the grease in the horizontal pipe or tubing becomes a grease plug that will not allow dirt or water vapors to reach the bearing.
- Do not rotate the motor. Do not regrease the motor, regardless of storage duration.

https://doi.org/10.1515/9783110749441-014

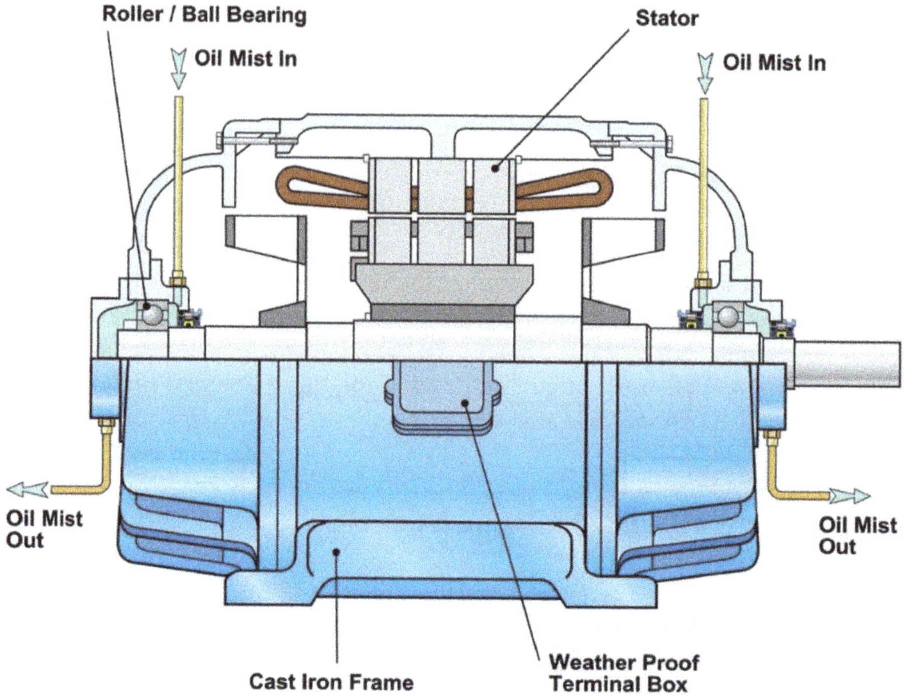

Fig. 14.1: Small motors are almost always provided with rolling element bearings, Oil mist is introduced between the magnetic seals and their adjacent bearings (source: AESSEAL Inc., Rotherham, UK; and Rockford, TN).

- Seal the shaft openings with silicone rubber caulking. Choose a dark color, like black to discourage pilfering.
- Coat all exposed machine surfaces with product C.

14.2.1 Relating bearing construction to "leave alone" strategies

As stated in the guidelines in the previous section, electric motors should not be given periodic regreasing while in storage, regardless of storage duration. The reasons for this experience-based statement become clearer when looking at typical ball bearings in their respective housings. Six of these bearings were shown in Part A of this book as Figs. 4.3–4.8. Suffice it to say here, they differ from each other in various ways, and incorrect regreasing can do damage. Therefore, giving the bearings uniform storage protection by sealing them off (and *not* first filling them with grease) is a sensible preservation strategy. However, such sealing off (done by simply leaving the motor and its bearings alone, except for placing a bead of caulk around the shaft opening at the motor end plate) is not as trustworthy as oil mist blanketing. With blanketing, oil mist reaches and protects every interior surface of the bearing housing. Oil mist intruding into the motor windings is of little consequence.

An examination of proper motor lubrication illustrates that, unless the same style of bearing is used on all the plant's electric motors, there is no substitute for following a proper work execution procedure. This procedure will be different for different styles of bearings. Good supervision and experience-backed lubrication management prevent failures and generate higher profits. But, again, for storage preservation of electric motor bearings in situations where oil mist is not palatable for reasons we cannot really fathom, try to follow the recommendations in conjunction with Fig. 14.1.

14.3 Large electric motors

Assuming the large motor has sleeve bearings, the following procedures should be followed for "old style"–meaning traditional as opposed to state-of-art– preservation:

1. Blank (i.e., insert blind) the oil return line.
2. Seal the shaft openings with silicone rubber caulking and tape.
3. Fill the bearing housing with product B.
4. Install a valved standpipe so that the inlet is higher than the bearing's housing.
5. Coat all exposed machine parts with product C.
6. Do not rotate the motor shaft.

14.4 Steam turbines

For steam turbines, the "old style" preservation process would be to:

1. Isolate the turbine from the steam system.
2. Seal the shaft openings with silicone rubber caulking. Select a dark color, like black, to discourage pilfering. Then, apply tape to the shaft's surface.
3. Dry out with instrument-grade (i.e., −40 °C dew point) air.
4. Fill the turbine casing, including the steam chest, with oil containing 5% rust prevention concentrate. Hold the governor valve open as necessary to ensure the chest is full. Vent the casing, as required, to remove trapped air. Fill the T/T (trip-and-throttle) valve with oil.
5. Install a valved pipe on the casing, which will serve as a filler pipe for adding oil to the casing. Allow space for thermal expansion of oil in the pipe.
6. Coat all external machine surfaces, cams, shafts, levers, and valve stems with product C.
7. Coat the space between the case and shaft protrusion with product A. Cover the space with black tape.
8. Fill the bearing housing with oil.
9. Coat the casing bolts with product C.

14.5 Gas turbines and hot gas turboexpanders

All manufacturers of gas turbines have developed and will issue specific and highly detailed storage preservation procedures. Following these procedures is necessary so as not to void the manufacturers' warranties.

14.6 Gearboxes

For gearboxes, the "old style" storage preservation procedures are:
1. Fill the gearbox and piping with oil containing product B.
2. Plug all vents and allow space for thermal expansion.
3. Apply product C to the shafts and couplings.
4. Tape the locations where the shaft protrudes through the casing or housing.
5. Install a valved (meaning a valve placed on top of a vertical pipe) pipe on the casing, which serves as a filler pipe for adding oil to the gear case.
6. On gear speed reducers, manually turn the input shaft a sufficient number of turns to allow the output shaft to make least three full turns. Do this every 3 months and also retape all locations where the shafts protrude.
7. On gear speed increasers, manually turn the output shaft a sufficient number of turns to allow the input shaft to make at least three full turns. Do this every 3 months and also re-tape all locations where the shafts protrude.

14.7 Centrifugal (dynamic) plant air compressors and blowers

The same "old style" preservation procedure applies to centrifugal compressors, dynamic plant air compressors and blowers. The steps are:
1. Purge the compressor's casing of hydrocarbons, unless dealing with a helium or other inert gas compressor.
2. Flush the unit's internals with solvent to remove heavy polymers.
3. Consider pressurizing the casing with nitrogen or oil mist. If this is not possible, continue with next steps, 4 through 11.
4. Fill the casing and all bearing housings with product A. Circulate product A through the entire system for 1 h.
5. Blank or blind oil return the header.
6. Seal the shaft openings with silicone rubber caulking and then tape.
7. Fill the compressor's bearing housing with product A and run the turbine-driven oil pump at a reduced speed for at least 1 h.
8. Fill the oil console with product A.
9. Fill the compressor with nitrogen when it is at an ambient temperature. Turn off all heat tracers.

10. Coat all exposed machine parts, including couplings, with product C.
11. Tape all shaft protrusions. Repeat this step after periodically turning the shafts.

14.8 Lube and seal oil consoles and circulating oil systems

The following "old style" preservation procedures should be implemented for lube and seal oil consoles, as well as circulating oil systems.

1. Install a filler pipe in the reservoir vent or replace the breather with a standpipe. Then, fill the entire reservoir volume with product A. Affix an elbow to the top of the filler pipe to prevent birds from entering and place, fasten and secure a transparent, wide mouth plastic bottle over the opening.
2. Circulate product A throughout the piping system. Open and close the control and bypass valves so the oil will reach and coat all piping and components. Circulate for 1 h. Vent the trapped air from all components and high points.
3. Block in filters and coolers. Fill with product A and allow a small space for thermal expansion. The water side of coolers should be drained and air-dried. Plug all vents. Lock all drain connections in a slightly open position.
4. Fill the reservoir or tank with product A. Blind or plug all connections to the reservoir or tank, including the vent stack.
5. Coat any exposed shaft surfaces and couplings of lube oil pumps with product C.
6. Tape any exposed shaft surfaces on the lube and seal oil pumps.

14.9 Reciprocating compressors

The "old style" preservation process for reciprocating compressors is as follows:

1. Purge all compressor cylinders of hydrocarbons.
2. Blank or blind the compressor suction and discharge.
3. Fill the crankcase, cooling water jacket, and valves with product B. Install a valved standpipe, allowing space for thermal expansion.
4. Coat all exposed machine parts with product C.
5. Top-up the level of product B in the cooling water jacket.

14.10 Hydraulic units

When using the "old style" preservation method for hydraulic units, follow these steps:

1. Fill the hydraulic oil reservoir with product A.
2. Circulate product A throughout the piping system. Open and close the control and bypass valves so product A will reach and coat all piping and components.

Circulate for at least 1 h. Vent the trapped air from all actuator components and high points.

3. Block in filters and coolers, if separate. Fill these with product A, but allow a small void space for thermal expansion.
4. Drain the water side of coolers, then air-dry. Plug all the vents. Lock the drain connections in a slightly open position and indicate this on a tag. Affix the tag to the drain valve.
5. Ascertain that the reservoir is filled with product A. Blind or plug all connections to the tank, including the vent stack.
6. Coat all exposed shaft surfaces and couplings of externally mounted hydraulic oil pumps with product C.

Proven practices are outlined above. These „old style" practices are rarely, if ever, the most desirable. Not following oil mist-related best practices may deprive commercial entities from reaching their potential.

Chapter 15
Strategy for short-term equipment storage preservation

15.1 Management digest

The probability of equipment failure is relatively high following the commissioning of equipment, whether for initial operation or after completing rebuild activities. Machines are often sitting idle at a new industrial site for months while site construction progress slows down. It is not unusual for projects to be canceled with no thought given to equipment preservation.

Unless the buyer clearly specifies storage measures in the contract, it is likely that machines will be shipped without provisions for storage. On-site storage protection needed for 3–12 months at a construction site requires a preventive maintenance (PM) program. At best-in-class performers, on-site storage preservation with oil mist is standard practice, and its cost is included in the initial budget for projects.

Only in dire circumstances would best-in-class performers revert to the conventional ways of temporary storage protection. This section highlights those circumstances. The strategies and recommendations refer strictly to short-term preservation using conventional, old, non-optimized protection methods. These plans and methods are described here for plants located in a northern, dry climate.

15.2 Shaft rotation requirements (applicable to short-term equipment storage)

Rotate all equipment, such as motors, turbines, gears, and compressors, at a minimum once per month, but optimally every 2 weeks. Keep in mind that for rolling element bearings, a single turn of the shaft means the bearing cage would have made only a half-turn. Therefore, rotate the shafts a minimum of two-and-a-half turns.

For gears, refer to the upcoming *Gears* section for the required number of shaft turns. On gears with teeth dipping in oil, you want to have each tooth dipped in the oil sump at least once each month. The number of recommended shaft rotations (i.e., turns) differs for gear speed increasers versus gear speed reducers.

Gear units where oil is pumped into bearings and sprayed into gear teeth must have their breathers plugged and their shaft protrusions caulked with silicone caulk and then taped. The void space above the oil level must be filled with oil mist or blanketed with nitrogen. No periodic turning of shafts is needed.

A tag explaining the protection and preservation method must be affixed to the gear unit, machine, compressor, or whatever the equipment may be.

https://doi.org/10.1515/9783110749441-015

15.2.1 Visual inspection (refers only to short-term equipment storage)

When rotating exposed machine surfaces, check the shafts and couplings to verify a protective coating of product C has been applied and that neither this protective coating nor the tape has been removed. Reapply both, if needed.

Check all lubricating lines to make sure no tubing, piping, tank, or sump covers have been accidently removed. Retape all ends and covers. If flanges are open on the machinery, notify the pipefitter, first-line supervisors, or other designated personnel.

15.2.2 Draining of condensate (for short-term equipment storage)

Drain the condensation from all bearing housings, sumps, and oil reservoirs on a once-per-month schedule. If an excessive amount of condensation is found, recheck once a week or at 2-week intervals, depending on the amount of condensate present.

15.3 Bearings (for short-term equipment storage)

Fill all bearing housings that are oil lubricated (and not force-fed) with rust prevention concentrate, bringing the oil level up to the bottom of the shaft. For force-fed bearings, the upper bearing cap and bearing must be removed. A coat of heavy, inhibited oil can be applied to the journal and bearing surfaces. This should be reapplied as needed.

15.4 Electric motors (for short-term equipment storage)

Electric motors with greased bearings should not be lubricated. If received with a grease fitting, it should be removed and the opening plugged or capped.

15.5 Steam turbines (for short-term equipment storage)

Spot-check turbines by removing the upper half of the turbine case and visually inspect. Plan to open a sampling of these turbines, selecting from the first preserved and those in the worst condition. This should be done on a 3-month schedule. Other turbines may be inspected by the manufacturer's field service engineer during monthly visits. This inspection is done through the opening in the top case while the rotor is being rotated on a 1-month schedule.

15.6 Gears (for short-term equipment storage)

Fog the interior of the housings with rust prevention concentrate. Coat all tooth contact points with an inhibited grease or heavy, tacky oil (such as product C). Remove the inspection plates and visually inspect the interior on a 3-month schedule. If no longer than 3 months storage, rotation of shaft will not be needed.

15.7 Compressors (for short-term equipment storage)

Applying atomized lube oil from fog nozzles is occasionally done. It is considered a less effective substitute for oil mist, but some equipment manufacturers recommend fogging "because we've always done it this way." It can be inferred that short-term storage preservation by conventional means often involves the compressor manufacturer. The manufacturer's representative should inspect the compressor(s) during monthly visits. The needed preservatives are applied under the representative's supervision; this will protect the warranty provisions.

Apply oil fog to centrifugal compressors' internals and consider placing desiccant bags in these machines. Ideally, inspect these compressors once a month or at least on a 2-month schedule. Inspect high-speed air compressors on a 3-month schedule. Inspect and fog axial compressors on a 3-month schedule.

15.8 Using oil mist for short-term equipment preservation

Oil mist is ideal for storage preservation mainly for its simple operating principles and excellent reliability. Among the reasons why oil mist is a beneficial method for short-term equipment preservation are:

- Really simple and straightforward oil mist lubricant applications
- Competitive costs for oil mist preservation if new consoles are chosen
- Reduced costs for oil mist preservation if the oil mist generator equipment is purchased on the used equipment market
- Very few man-hours are typically expended for its maintenance or surveillance
- Oil mist lubrication and oil mist preservation are identical except for oil consumption. Oil mist for standstill equipment preservation consumes only 10% of the amount consumed for oil mist lubrication.

The importance of oil mist for short-term equipment preservation can be reviewed by revisiting one of the earlier chapters of the book.

15.8.1 Other considerations for short-term equipment storage

In a warm, high precipitation climate, it is best to look for optimized solutions for field storage during construction and prior to start-up. If oil mist lubrication is not already part of the original design, it should be seriously considered because, in the overwhelming majority of cases, it will provide the best protection against contaminant ingress.

Oil mist will reach all spaces and voids. Other methods may not reach oil galleys and condensation hanging in droplet form from casing covers and other parts.

When looking at the cost outlays for routine maintenance and the related activities needed with conventional equipment storage preservation, many are surprised that these can greatly exceed the outlays for oil mist preservation.

15.8.2 Storage preservation mentioned in industry standards

A number of equipment standards and industry guidelines exist on the topic of storage protection. Among them are references in API documents:

- API-RP-686 3.2.1 states: "When more than ten pieces of equipment are to be stored for a period of longer than six months from time of shipment, oil mist protection should be considered."
- API-RP-686 3.2.2 notes: Oil mist should be used to protect the bearing housings, seals areas, and process end of equipment.
- API-RP-686 3.2.3–3.2.16 state the full API recommendations.

The manpower needs for storage and preservation would be much higher for vapor phase inhibitor application (e.g., preparation for storage and preparation for installation and commissioning) as opposed to oil mist preservation. The latter would leave the equipment, including any machine used in the petrochemical, oil refining, and mining industries, ready for installation and start-up.

Documented accounts are available where oil mist has, in past decades, provided reliable protection for idle and standby equipment at many facilities. Project executives seeking on-time commissioning and high future operating reliability are encouraged to provide the same well-proven protection for current projects and its new equipment during storage.

To date, no facility has ever published the embarrassing statistics of early bearing and mechanical seal failures encountered when trying to omit equipment storage preservation. However, infant mortality, defined as failures within 30 days of start-up, is exactly what has been observed by large and small companies that temporarily idled plants or units with inadequate or no effort spent on storage protection (i.e., mothballing).

One 2013 incident involved a facility that had initially "saved" a few dollars by not using the best equipment protection. Those in charge later tried to get advice on "what could we do now" from consultants. Risk reduction strategies were mapped out, which would cost the facility millions in undoing the damage. Whenever machines are installed with in-place spares, one of the two is to be dismantled by contract personnel. Company retirees were asked to supervise the machine dismantling and reassembling tasks. Decisions relating to work on the second machine were made with the full knowledge gained from the first dismantling and reassembling.

This situation would have been less costly if the reliability professionals at the facility had aimed for the implementation of best available technology. Reliability professionals who do not see this as their role have no business calling themselves reliability professionals. At all times, they would do well to remind themselves of the principles of equipment storage preservation:

- Operating machinery must be protected from the elements. Painting, plating, sheltering, use of corrosion resistant materials of construction and many other means are available to achieve the desired protection.
- A similar set of protection requirements applies to not yet commissioned or temporarily deactivated equipment.
- Storage preservation makes economic sense. However, the quest for safety and reliability must be first, saving money should be considered a close second.
- Early in the chain of decision-making, management deserves to be thoroughly briefed on best available technology.
- Bring in competent advisers early on in the process. Regrettably, competent advisers are often brought in *after* the damage is done. On average, incompetent opinionators are often involved *initially*. They are holding on to their jobs by always supporting the boss. Opinionators are rarely asked to explain why best in class competitors have made it corporate policy to budget and include oil mist preservation in their projects.
- Follow these principles for both new (i.e., not yet commissioned) and mothballed (i.e., previously operating but now shut down) or inactive machinery.
- The work processes or procedures chosen for the preservation or corrosion inhibiting of fully assembled but inactive (i.e., mothballed) process machinery will logically depend on the type of equipment, expected length of inactivity, geographic and environmental factors, and the amount of time allocated to restore the equipment to service.

The basic and primary requirement of storage preservation is the exclusion of water from those metal parts that would form corrosion by-products. Rust is one of the by-products of corrosion that could find its way into bearings and seals. A secondary requirement might be the exclusion of sand or similar abrasives from close-tolerance bearings or sealing surfaces. All or any of the storage preservation strategies must aim at satisfying these requirements.

Machinery preservation during pre-construction, storage or long-term deactivation (i.e., mothballing) will affect the machinery's infant mortality at the start-up of the plant or process unit. Many times, machinery arrives at the plant site long before it is ready to be installed at its permanent location. Unless the equipment is properly preserved, scheduled commissioning dates may be jeopardized and the risk of failure would increase.

15.8.3 Protection of mechanical seal components in nonoperating fluid machines

Protecting liquid or dry gas types of mechanical seals in pumps that are stored or mothballed for prolonged periods of time deserves attention. The precautionary recommendations of at least one mechanical seal manufacturer are provided as a general guide.

The guidelines for mechanical seal and bearing storage are straightforward. When possible, remove the mechanical seal from the pump. Mechanical seals are assembled and tested in a clean room environment. Left unprotected in the field, both internal and external contaminants, such as airborne dust, can accumulate in the critical sealing areas, causing leakage or damage to the seal upon start-up. This may equally be true for liquid and dry gas seals.

The seal should be labeled to identify its materials of construction, then packaged and stored in a controlled environment. A consumer-type vacuum sealing apparatus can be used to preserve and protect the seal. If the seal has been in operation previously or has been in contact with fluids, it should be returned to the manufacturer for inspection and/or repair.

Bearings in storage in a warehouse should remain wrapped in their factory plastic, wax paper, or vacuum pack. They must be laid down flat, not standing on edge.

15.8.4 Pumps and fluid machines where no fluid is present

Whenever possible, process fluid is drained before applying the storage or protective treatment. In that case, the guidelines are:

- Valve off the pump and drain the fluid from the pump casing.
- Remove the seal's environmental controls and plug the ports.
- Drain all other fluids, including the barrier fluid, from the seal.
- Clean the seal chamber(s) with a solvent that is compatible with the seal's materials to remove all possible residues. Next, verify that all fluid/solvent has been drained from both the pump and seal. Rotate the shaft by hand during this process.
- Plug all seal ports.

- Mask the opening between the shaft or seal sleeve and the gland to protect the seal from environmental contamination.
- Tag the equipment with the date of storage.

15.8.5 Pumps and fluid machines where fluid is present

In these instances, the recommended general guidelines are:
- Remove the seal's environmental controls.
- Plug all seal ports.
- Securely mask the opening between the shaft or seal sleeve and the gland to protect the seal from environmental contamination.
- Tag the equipment with the date of storage.

15.9 Case history involving EPC contractor

Engineering, procurement, and construction contractors (EPCs) occasionally ask about readily accessible literature that discusses practical steps for switching back and forth from preservation to operating modes when using oil mist. Still, other EPCs often pass on requests from clients.

As an example, some user-clients have asked for leakage calculations when dry nitrogen is used as a purge or sweep system on a condensing steam turbine. In the process of educating themselves on the various aspects of nitrogen purge preservation systems, one interested EPC came across older articles on the use of oil mist for machinery preservation. Although not directly involved with any meetings with his client, the EPC wanted to know what alternatives the client or potential user may have. The EPC anticipated a future involvement and wanted to be able to discuss it, if called upon to do so.

The EPC consultant was trying to understand the practical aspects and relative merits of nitrogen purge versus oil mist preservation systems, especially with regard to condensing steam turbines. Since one of his client's plants involved an installed turbine and a surface condenser, the issue of water removal prior to introducing the oil mist and then the removal or cleanup of the oil prior to starting the turbine was something he wished to explore. He wrote:

> As things proceed with this project, I expect to get the opportunity to ask some questions about whether the client has considered oil mist as an alternative to nitrogen purge. At that time, it will be very helpful to be able to mention that others have much experience in this field, and also the fact that the largest U.S. oil refineries have found pure oil mist to be a more economical way to go. Although I will not be making the decisions about which services or equipment to purchase, I will be sure to mention these refineries as experienced resources in this area.

The EPC engineer quite correctly noted that there is an *incremental initial* cost involved in utilizing oil mist preservation for grassroots projects. To his uninformed clients, it will always seem expensive and perhaps move the decision toward vapor phase inhibitor (VPI) products. However, oil mist preservation maintenance outlays are least cost; added cost is always involved in other storage methods because personnel must keep the bearing housings filled with the inhibitor. They will have to rotate equipment shafts and reseal the location where shafts protrude from casings. In the event that shafts are not periodically rotated, bearings will tend to develop flat spots.

No matter the environment, be it hot, dusty, damp, windblown, or cold, and with the long-term storage needs being longer than 6 months with many projects, oil mist preservation is the right choice to ensure low or no infant mortality of new equipment. Not only does oil mist preserve day and night in all possible weather conditions, it also helps to prevent flat spots on bearings caused by ground vibration. The oil mist completely coats the bearings and provides an oil film to cushion the bearings from vibration.

We sent the EPC consultant a synopsis of relevant portions of this text. In particular, we pointed out that reasonable cost projections using pre-owned oil mist equipment were available (see earlier chapters). We assumed that with 3 years of straightforward oil mist protection at most 6 of 100 asset samples were expected to experience infant mortality. In contrast, we anticipated that fully 100 assets would have to be dismantled and rebuilt if left unprotected for 3 years.

There is another way of explaining field experience dating back two or three decades: Without dismantling and rebuilding, improperly maintained assets using traditional preservation are a threat to operating reliably. Not performing maintenance or leaving the stored equipment unprotected will escalate the number of assets succumbing to infant mortality. An unacceptably high number of equipment failures would occur during the first year after startup of new equipment/units unless precautionary dismantling and rebuilding would be done. Based on our projections (see Chapter 11) at most two infant mortalities were expected for equipment re-started after a full year with oil mist. In contrast, 22 infant mortality incidents were expected with 100 assets left unprotected for a year. A compromise approach was implemented by a client who dismantled, cleaned, and refurbished one half of the assets for which installed spares existed.

Conventional and traditional equipment storage protection and preservation make use of grease-like coatings and/or vapor phase inhibitors. The manufacturers and providers of these products issue the type of information we find here.

Best practice is to compare different versions of the manufacturers' guideline and the guidelines found in books such as this. Reliability professionals then reconcile different versions, pointing out pluses and minuses. Oil mist usually wins these contests with little or no difficulty.

Chapter 16
Preparing stored equipment for re-commissioning (re-start after long periods of preservation)

16.1 Management digest

Suppose oil mist preservation had been applied to a machine in an oil mist storage yard. Whenever that machine is being moved to its permanent installation site (usually a foundation), turn off the oil mist system or simply remove the plastic oil mist tubing at the manifold. Plug (or cap) the manifold opening from which the tubing has been removed.

Regardless of how many plastic tubes are being removed from the preservation system and how many manifold openings are being plugged, the total volumetric output of an oil mist generator remains unchanged. Because one fewer asset is now connected to the system, the oil mist header pressure will increase until the oil mist cabinet controls are reset to achieve the pressure reading it had before.

16.2 Steps before removing machine

Before a machine is removed from the oil mist storage yard, remove the casing drain plug and allow the coalesced oil to flow out. With the possible exception of steam turbines, where residual oil might contaminate the steam condensate recovery system, the bearing housings and, occasionally, the interiors of all oil mist-protected machines can now be refilled with whatever lubricant was recommended for equipment in operation. No commercial situations are known where the thin oil mist film clinging to interior parts or surfaces needs to be removed.

Product C, which had been previously applied to the exterior surfaces of all machines, can now be removed with a solvent recommended by the product's vendor or supplier-manufacturer.

If a project team uses the guidance in this book, the interior spaces of the various machines may perhaps have been coated or filled with product A or B. Both are inhibited turbine oils that can, in some cases, stay in the equipment for reuse while operating. In other cases, the machines were well protected so that they can operate at part load or at normal load for a few days.

If the decision is made to drain product A or B while a machine is in the storage yard, the equipment can be tilted. This allows the preservative to flow to the low point drain and the drain plug is then removed. After the solid drain plug is again threaded into the drain port, the machine is ready for its normal refill. Oil quantity and type should correspond to the manufacturer's recommendations.

Be sure to verify that the breathers are back in place. There also may be electric heaters or instruments that must be reconnected.

https://doi.org/10.1515/9783110749441-016

16.2.1 The process pump example

Using process pumps as an example to guide our thinking, we might dwell briefly on a typical cleaning sequence. Recall an earlier chapter where we had described the straightforward steps involved with removing a fluid machine (pump) from an oil mist preservation yard.

For machines that had been in storage while protected with coatings, we would simply follow the preparatory sequence below. In essence we would be preparing to restart after donning protective clothing and using solvent to remove the external protective coating, usually product C:

- Check with the seal manufacturer to verify that the date of storage, in conjunction with the materials of construction, does not exceed the shelf life of the mechanical seal. If the expiration date is exceeded, check the face condition to ensure flatness.
- Remove the masking tape, duct tape, or sealant from the opening between the shaft or seal sleeve and the gland.
- Valve off the pump and remove the plugs from the seal ports.
- Flush with a solvent compatible with the seal's materials to remove all possible residues. Rotate the shaft by hand during this process.
- Drain the fluid from the casing.
- Reconnect the seal's environmental controls and/or plug the seal's ports.
- Open the suction valve fully and crack open the discharge valve(s) about 10% of full travel.
- Vent the seal chamber to allow the seal to become surrounded by the liquid.
- Start the pump, using a relevant, approved, and pump-specific checklist or procedure.

16.2.2 Inert gas purge versus the oil mist preference

Unless defined otherwise, disconnecting the inert gas line and/or oil mist supply line is all that is needed to prepare a machine or asset for removal from the outdoor storage yard and permanent installation on site. No steam purge will be needed.

The same preparatory sequence can be followed if vapor phase inhibitor products have been used. However, if for any reason their residue must be removed, steam cleaning may have to be used just prior to on site commissioning.

Best practice is for a responsible reliability professional to discuss the above guidance with the crews given the task of implementing the recommended procedures. Obtain their buy-in and/or resolve differences before instructing the implementers to proceed.

(with involvement in equipment-upgrading work). What LL (not his real initials) told us is rather telling.

LL knew that original equipment manufacturers (OEMs) want an end-user/purchaser's spare parts business. In his email, he described attending a meeting with a large-scale-equipment end-user, the user's engineering contractor, and a representative of a major lobe-blower OEM. The end-user and engineering contractor both wanted pure mist on the bearing end, and purge mist (liquid oil with oil mist blanketing) on the timing gear end of the lobe blower. The OEM's representative, however, refused to approve pure oil mist and warranty it. Even more interesting, though, is the fact that he told everybody in the meeting why: He said that he would not be able to sell as many bearings and parts for his lobe blowers if he approved the installation and use of pure oil mist.

17.4 Be mindful of the bottom line

When manufacturers of plant equipment have no interest in upgrading the reliability of their products, it will be up to the end-user or owner-purchaser to demand better performance. Surely, solid, well-detailed end-user or owner's specifications that include purchasing and upgrade measures adopted by best-practices facilities would do wonders here. But unless a corporation or facility has such specifications, it could find itself dealing with narrow-focus and repair-intensive responses from vendors/suppliers similar to that offered by the representative of the lobe-blower-OEM described in LL's email.

As the final choice of a product offering is clearly up to the buyer, end-users or owner-purchasers must educate themselves. Unless they insist on quality, they will often receive maintenance or repair-intensive products. Relying entirely on engineering/procurement/construction contractors and OEMs is simply not enough. In the final analysis, the end-users or owner-purchasers get what they deserve: Either more downtime risk and bloated budgets, or higher equipment reliability and the profits that come with safely extended uptime.

So, what would you and others at your site choose? For greater safety, asset reliability, and profits, consider placing this book in the hands of your fellow employees. And if you are a manager who periodically conducts employee performance reviews, the book will help you ask questions and judge the validity of the answers that flow in your direction.

Appendix I: Damage terms, damage prevention, and the corrosion mechanism

Normally, two types of unexpected damage can harm inactive equipment: false brinelling and static corrosion.

A.1.1 False brinelling

False brinelling occurs when the lubricant is squeezed from between rolling elements and a raceway due to vibration, not rotation. It often occurs during storage or transport when minute vibrations in the environment cause the rolling elements to oscillate slightly. The angular motion from the vibration is too small to generate or support a full lubricant film. Metal-to-metal contact then results, tearing microparticles from high points on the surface.

False brinelling appears as bright, polished depressions in the raceway beneath the vibrating elements. It also may have the characteristic red-brown stain of fretting. Oxidation at the point of contact determines the appearance. If there is a slow oxidation rate, false brinelling depressions remain bright and are often confused with true brinelling. True brinelling occurs when there is instantaneous excessive pressure, such as a hammer blow to a cage during installation. With true brinelling, metal is displaced away from, and symmetrically surrounding, the point of contact.

The spacings of false brinelling marks on a raceway are equal to the distance between the rolling elements, just as it is in true brinelling. If bearings are rotated slightly between periods of vibration, more than one pattern of false brinelling may be seen.

False brinelling damage can be observed even though the forces applied during vibration are much smaller than the static carrying capacity of the bearing. However, the damage becomes more extensive as the contact load on the rolling element bearing increases.

Three conditions are necessary for false brinelling to occur:

1. The bearing must be under load, either from the weight of the shaft or the weight of the bearing itself.
2. The bearing must not rotate.
3. There must be an external source of vibration.

Most inactive or standby industrial equipment is subject to these conditions.

https://doi.org/10.1515/9783110749441-018

A.1.2 Preventing false brinelling

There are several ways to prevent false brinelling. We will take a look at each one.

- **Minimize the load by blocking.** Weight may be lifted off the bearing by blocking the shaft. Blocking consists of supporting the shaft at both ends by an external support frame (think sawhorses) or block. Blocking the shaft protects electric motor bearings indefinitely against false brinelling. Vertical equipment may also be blocked to take the thrust load off the supporting bearing.
- **Rotate bearings with a periodic rollover.** No matter where the equipment is stored, the bearings should be rotated at least monthly to ensure that the lubricant film covers the bearing's contact area. If there is equipment operating nearby that generates vibration, especially in warm, moist environments, once a week is preferable.

Ideally, rotating equipment assemblies should be turned two and a half turns. Because the bearing cage makes one full turn for every two turns of the shaft, making fewer than two turns will probably not do the job. While this now redistributes lubricant evenly and ensures that the balls, rollers, gears, or shaft are not in the original position, ascertain that the number of shaft rotations harmonizes with the tooth ratio or number of gear teeth. The intent is to "wet" all gear teeth. On epicyclic gearing ask the manufacturer to advise how many turns of the input or output shafts are needed to fully wet each tooth.

- **Grease and rotate lubricated components.** Lubricated components also should be periodically rotated. With the possible exception of lifetime lubricated bearings, components should be inspected and regreased annually. Product C may be needed on external surfaces for extended protection against rust and corrosion.
- **Visually inspect during rollover.** Check the shafts and couplings during rotation to ensure that the rust protection film is still viable. Reapply if needed.
- Check all lubricating lines to determine if any tubing, piping, tank, or sump covers have been removed. If a vapor space inhibitor product is used, make sure the equipment is resealed after shaft rotation. If necessary, re-tape the ends and replace the cover.
- Inspect the interior of lube oil consoles. Check to see if the reservoir is clean and free of rust and condensate. Clean and dry if needed, then fog with rust prevention concentrate. Unless proper nitrogen blanketing or oil mist are present, be aware of rust icicles forming on the inside covers of oil reservoirs. Think oil mist preservation and implement it on oil reservoirs.
- **Intermittent operation.** Running the machine periodically keeps seals lubricated, prevents them from taking a set, and minimizes vibration or corrosion damage.
- **Move away from sources of vibration.** Some environmental vibration is to be expected, especially if the equipment is close to operating machinery, roads, or railroads. Consequently, the equipment should be stored as far away as possible from external sources of vibration.

A.1.3 Static corrosion, its mechanism, and prevention

Static corrosion or rust is an electrochemical reaction that occurs when water comes into contact with a metal surface. Rust formation can be accelerated by the presence of various contaminants in the electrolyte or on the metal surface. Keep away scale, dust, acid, salt, or alkali.

To produce rust, the following conditions must be present:

- An anode (hills)
- A cathode (valleys)
- An electrolyte (water)
- A complete circuit (bearing/gear surface)

The anode is the area where corrosion takes place. The electrolyte must have the ability to conduct a current, regardless of how small. Also, the electrolyte must be in contact with the anode and cathode of the metallic surface. The circuit is completed by the metal surfaces in bearings and gears that are likely in the path between the anode and the cathode.

Because water contains hydrogen, hydroxyl ions, and, usually, dissolved oxygen, a molecular film of moisture on the metal surface sets up the following reactions:

Reaction at anode:

$$2Fe \rightarrow 2Fe^{2+} + 4 \text{ electrons}$$

Reaction at cathode:

$$4Fe^{2+} + 8OH^{-} \rightarrow 4Fe(OH)_2 + 2H_2O \rightarrow 2Fe_2O_3 + 4H_2O + H_2$$

Now we have the two components needed to form the very temporary intermediate product $Fe(OH)_2$. Rust, Fe_2O_3, will now form:

$$4Fe^{2+} + 8OH^{-} \rightarrow 4Fe(OH)_2 + 2H_2O \rightarrow 2Fe_2O_3 + 4H_2O + H_2$$

The anode reaction, in addition to producing red or orange $Fe_2(OH)_3$, can also produce, depending on the amount of dissolved oxygen available, black FeO or Fe_3O_4. This reaction occurs very rapidly on the steel surfaces of bearings. The oxides of iron, which are often harder than the base steel, act as abrasive lapping compounds when the bearing is put back into service. The resulting pits cause concentrations of stress that reduce the load carrying capacity of the bearing and gears, causing premature wear and failure.

To prevent static corrosion (i.e., rust) from occurring in inactive equipment, a major lubricant manufacturer recommends one of the following four storage procedures: Fully lubricated storage, semi-dry storage, dry storage, and oil mist lubrication.

A lubricant manufacturer's short instructions often consist of brief overview sheets describing categories of storage protection. We re-state these categories to allow responsible individuals to make a side-by-side comparison and supplement whichever one of three available primary formats they choose to follow. Since semi-dry is a simple composite of dry and fully lubricated storage, the choice is concentrating on fully lubricated, dry, and oil mist approaches to storage preservation.

A.1.4 Fully lubricated storage

If practical, leave the circulating lube system heated, in operation, and with oil flowing through the bearings and gear cases to:

- Keep bearings flushed of condensate or water
- Keep drain drop legs clear of cold oil
- Keep drain headers warm and flush water and debris flowing back to traps and reservoir
- Keep seals wetted
- Keep oil warm in reservoir to break out collected water
- Keep oil from precipitating and shortening filter element life
- Keep flow meter seals wetted and sight flow pistons from sticking
- Keep gear cases protected from corrosion

A.1.5 Semi-dry storage

This procedure can be used for temporary or short-term storage. Drain, flush, and clean the reservoirs. Spray the equipment's interior components with enough of a good VCI product to cover the bottom of the reservoir and lightly coat exposed surfaces. The equipment owner should be aware that careful reapplication of the VCI product is usually required every 6 months. Pay close attention to shafts being turned. Rotation may break the seal and require resealing and re-taping.

A.1.6 Dry storage

This is another temporary or short-term storage option. Drain, flush, and clean the reservoirs. Spray the interior components of the equipment with a light coating of product A. Ensure all surfaces have been coated. Use product C on external surfaces. Greased components should be stored.

A.1.7 Oil mist lubrication

As a brief reminder: Oil mist coalesces on bearings and metal surfaces, thereby keeping them coated with oil. The slight positive pressure (0.1–0.3 psi above ambient) of the oil mist inside bearing housings and machine cavities prevents the ingress of external contaminants. Without positive internal pressure, bearing housings and machine cavities breathe with the changes of ambient temperatures. The in and out breathing cycle leads to moisture and construction related dust and dirt entering the equipment. Oil mist prevents this from happening.

The API (American Petroleum Institute) recognizes the effectiveness of oil mist machinery preservation in its recommended practices standard API RP-686. Section 3.2 states: "When more than ten pieces of equipment are to be stored for a period longer than six months from time of shipment, an oil mist protection system should be considered."

Lubrication technology has seen advancements. It follows that readers link up with lubricant providers who see fit to grow, groom, hire, and reward good application engineers. For these providers and their application-oriented support staff professional growth is a two-way street. An application engineer advising clients on bearing technology in a facility using or contemplating the use of oil mist would want to think of strengthening their bond with this client. Reading and considering [54–60] would, the author believes, be helpful and relevant. That said, the following subheading and associated narrative makes the powerful point.

A.1.8 How working with application engineers adds value

When it comes to optimizing equipment and storage, making the lubricant vendor-manufacturer your technology provider is a highly effective approach. Leading technology providers tend to employ application engineers. In case you're wondering what is needed to call oneself an "application engineer," here is a brief explanation.

The designation "application engineer" presumably originated decades ago in manufacturing companies producing rolling element bearings, mechanical seals, and similar products. These entities and their respective outlets deliver critically important components to both equipment makers and end users. In fact, application engineers working for lubricant providers are often known as cross-trained, multiskilled subject matter experts (SMEs), people with above average knowledge of the machines for which they sell products. These application engineers know where and how the lubricant is optimally dispensed, how clean it should be under ideal operating conditions, if and how these ideal conditions are realistically obtained and maintained in your machines, how oil can be kept clean, and when a charge of lubricating oil should be replaced.

Application engineers also develop application engineering summaries for vendors or suppliers to distribute to end users. Often, these summaries are presented in an eye-catching, yet comprehensive graphic format, as shown in Fig. A.1.1.

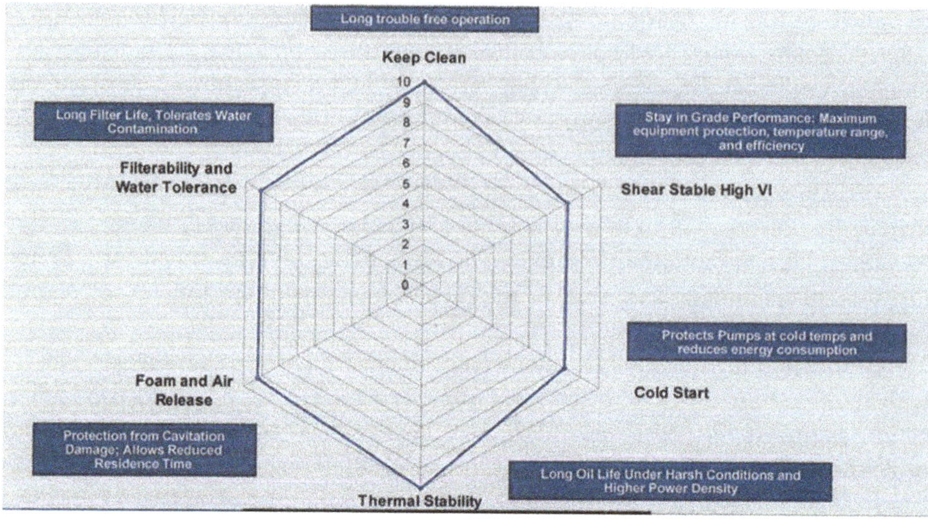

Fig. A.1.1: Spider diagrams are used by application engineers employed by top lube providers; they illustrate properties of a superior lubricant (source: ExxonMobil Lube Marketing Bulletin).

First and foremost, understanding the machines that need lubrication means knowing what makes them perform over extended time periods without excessive maintenance requirements. For application engineers, the knowledge ingredient in their work activities includes fluid machinery maintenance details and potential upgrade opportunities. Think about how this differs from salespeople who usually don't know about the labor-intensity of their proposed solution or the fact the product you're about to order will eat away the O-rings in your so-and-so machine.

Experienced application engineers can guide and track how well you, the client, are doing. Among other contributions, they can often compare a client-user with the client-user's competitors by assessing "us" versus "them."

For example, think of an application engineer in a rotor storage and preservation situation. Chances are you would expect this individual to understand the pros and cons of lifting the rotor off its support cradle so as not to cause excessive loads to act on just one contact point for 2 or 3 years. A good application engineer knows that large machines, if equipped with rolling element bearings, will develop fretting spots or false brinelling if subjected to vibration from an adjacent operating machine or railcars that travel back and forth on the embankment right behind your storage facility. One could expand on an application engineer's desirable attributes to say she or he are excellent teachers of reliability improvement, increased safety, and higher bottom-line profitability, among many other topics of potentially great value.

A real-life example involves a compressor application engineer who was employed by a manufacturer of positive displacement and dynamic process gas machines. After years as an installation technician and designer's assistant, he transferred to his company's sales division. Whenever he visited the offices of a multinational refining corporation to submit a compressor proposal, he combined his sealed bid submission with a free tutorial on compressor application matters. He brought along overhead projector transparencies (in 1960) that gave details related to application parameters of great interest to his target audiences. Young engineers and other recent hires at these offices signed up to attend these tutorial meetings. Word spread and more attendees were eager to sign up for the next opportunity to attend his application focused tutorials. The compressor manufacturer thrived during this time.

But when the compressor application engineer's new managers decided to discontinue this perceived free education of the future movers and shakers, the compressor manufacturer soon lost its competitive edge and became a company that no longer set itself apart from the competition. As the compressor application engineer's company shifted to compete on price alone, its quality of design and manufacturing began to suffer. Little did the company's managers know that one above average application engineer had made major contributions to the company's image, profitability, reputation, and success. Although difficult to quantify, disseminating his application know-how had greatly added to everybody's happiness in more ways than could be counted.

Thoughtful employers, and especially lubricant manufacturers, would likely spend money and effort on grooming and advancing their application engineers. Alternatively, the employer might consider hiring someone with at least a decade of experience in the business of fluid machinery, mining equipment, or whatever other machine categories use the manufacturer's products and would benefit from this engineer's knowledge. Teachable individuals, like a potential application engineer, are encouraged to trade some of their recreational time to advance their knowledge and technical education. Reading and learning are essential to becoming an application engineer, meaning an above average value-adder.

Certainly, an application engineer's employer should pay this key employee a salary commensurate with the person's acquired and demonstrated application experience. To the extent that application engineers advance to join the ranks of SMEs, the employer would advertise the prowess of its SMEs and why the seller's markup for a gallon of lubricating oil will probably exceed that of the less knowledgeable competition. The purely cost-focused competition caters to those who purchase from the lowest bidder, whereas the provider with application engineers and/or SMEs seeks out customers that aim for value. Buyers that put their trust in low-cost products often act on opinions, guesses, and out-of-context statements. These out-of-context statements may be true, but they rarely tell the whole story. Out of context statements very often apply to isolated and narrow cases, at best.

For instance, "I'm selling you an oil with a higher viscosity" is probably a fact. Expect a good application engineer to tell potential customers what benefits derive from using this high-viscosity oil and when, where, or why one would aim to avoid such oils. With this kind of authoritative input, the statement is placed in context and becomes a value-adding fact. Application engineers emphasize *adding value* over simply selling products. Salespeople, on the other hand, tend to prefer *selling product and making money* over adding value.

It should also be noted that most US states require a degree from an accredited college or university before using the designation "engineer." In all the US states, individuals must acquire a state license to call themselves professional engineers. Nevertheless, for many reasons, certification is an important step in the right direction. Studying for and then passing a rigorous test certainly shows the initiative and the desire to gain knowledge, although certification is not necessarily indicative of superior wisdom having been imparted to the certificate holder.

Appendix II: A new development: "ADIOS"

A.2.1 Status overview

As this book goes to press, Houston-based T.F. Hudgins is involved in a development that is slated to bring oil mist and/or oil spray to the individual machine or asset. But, before looking into this innovative development, keep in mind Tables 2.1 and 2.2. Recall that every manufacturer of rolling element bearings considers a narrow stream of liquid oil spray aimed directly at a bearing's rolling elements (the cage or ball separator) the *best available lubricant application*. Recall, also, that oil mist is ranked a very close second. There are no exceptions to these findings, and the correctness of this ranking has been verified on thousands of occasions, most of them empirically and some analytically. Now, suppose "ADIOS," an entirely air-driven intermittent oil spray, is built-in, or attached to the bearing housing of a machine. Suppose ADIOS is small and cost-effective, and that it will provide lubrication to equipment not conveniently reached by a plant-wide oil mist header system. In that case, ADIOS (Fig. A.2.1) represents a highly attractive solution to the facility's need to reliably lubricate up to four bearing housings or "points."

The ADIOS unit would be connected to an instrument air header or metallic air tubing. It would be fully pneumatic, and no electrically activated components would be involved. Constant level lubricators, oil rings, desiccant breathers, transparent sludge observation containers with petcock-like drains, and level observation ports (sight glasses) will have been discarded. Air would be vented through a coalescer; oil would be stripped from the air (from the oil mist) and fall back into the small add-on reservoir. If a machine or bearing housing so lubricated is large enough, ADIOS might even fit inside the bearing housing of such a machine. Rigorous calculations showed that very little oil would escape, and the oil supply would perhaps require topping-off a 2-gal (7.5 L) oil reservoir only once per year.

A.2.2 Oil rings have shortcomings

Why oil rings are still prevalent has much to do with initial cost and old traditions. Even today, many machines are still purchased with only first cost being considered. There is, thus, little (if any) incentive for manufacturers to compete on the basis of high quality and/or low maintenance requirements.

Chances are that the continued advocacy of oil rings involves just that, that is, a combination of low initial cost and tradition. Ever since 1781, when James Watt improved on Thomas Newcomen's 1712 steam engine, equipment manufacturers have provided oil rings as an inexpensive and generally accepted means of lubricating bearings in machines. But that was when plants employed a lubrication engineer and a designated oiler, and when oil was changed three or more times each year.

https://doi.org/10.1515/9783110749441-019

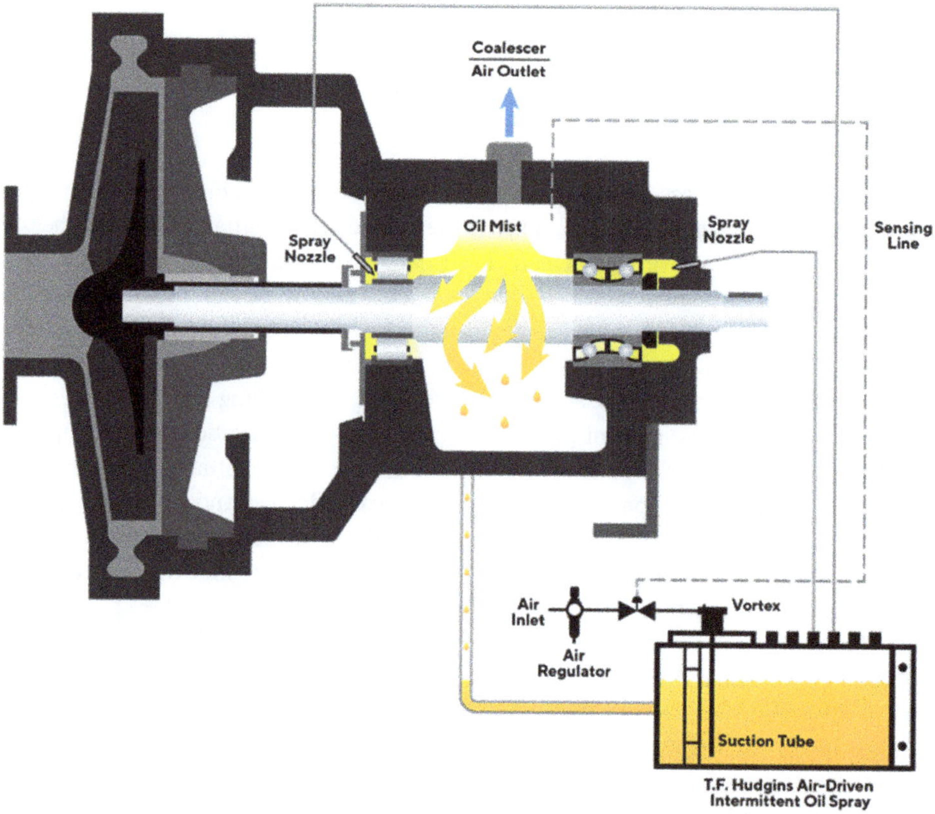

Fig. A.2.1: Schematic representation of ADIOS (air-driven intermittent oil supply) (source: T.F. Hudgins, Houston, TX).

That said, oil rings were, and indeed still are, acceptable components. Recall, however, that oil rings will meet today’s plant safety and reliability requirements *only* if manufacturers and users obey several very important requirements:

1. The shaft system must be horizontal. If the shaft is at even the slightest angle, an oil ring will travel to the downhill side and contact a stationary component. It will then slow down and may frequently abrade and contaminate the oil. Contaminated oil has a hugely detrimental effect on bearing life.
2. Oil rings must be concentric within 0.002″ (0.05 mm). To remain concentric within this tolerance, brass or bronze rings require manufacturing steps that include stress relief annealing after rough machining. Following rough machining and stress relief annealing, oil rings must be given chamfers or beveled edges and a final machining step. Cheap oil rings are quite obviously not manufactured with these sequential manufacturing steps.
3. Even the best available oil rings must be designed for optimum performance in a specific lubricant and within a narrow range of viscosities. The often practiced

"one-size-fits-all" approach will have disappointing results for the reliability-focused among us.
4. For proper functioning, an oil ring must be immersed in a lubricant to a specified or defined depth.
5. Oil rings do not provide lubrication at standstill, nor will they be of use for storage protection of equipment. When not running, a ring-oiled machine will usually ingest atmospheric moisture and airborne dust.
6. Constant level lubricators must be set to a specified height because they incorporate a transparent bottle and a bead of sealing caulk. As the caulk ages and goes through many hot and cold temperature cycles, it develops cracks. In an outdoor environment, rainwater can get past the caulk and reach the lubricating oil via capillary action. Therefore, even a properly height-adjusted constant level lubricator should be replaced on a time-based maintenance schedule.

For all six of these possible deviations *not* to exist or for all six to be adequately remedied is as improbable as discovering orange groves above the tree line in the Rocky Mountains. According to bearing manufacturer SKF, 91% of all rolling element bearings fail prematurely. The six reasons explain some of the factors that likely contribute to lube-related distress events in general purpose machines at many plants.

A.2.3 History and operating principles

Suppose leading pump users are risk-averse and others join them in the future. Consider an average size refinery with 2,000 process pumps. Even if the refinery has as few as 100 burdensome pump bearing failures per year, it may be worth considering the non-electrical, no battery power, fully pneumatic, ADIOS solution. Installing ADIOS would allow an oil refinery or other industrial user to eliminate oil rings, constant level lubricators, desiccant breathers, water observation ports, and frequent lube oil replacements.

Many bearing failures will disappear with ADIOS, an oil spray or oil mist delivery method that was originally designed for a manufacturer of mechanical seals. The manufacturer (AESSEAL) retained the right to use ADIOS for potential innovative methods of cooling and lubricating mechanical seals in the future. The consulting engineer whose original contributions proved helpful was authorized to fully disclose the underlying principles of physics, hydraulics, and pneumatics to a competent entity desiring to commercialize the technology.

As do all other oil mist systems, ADIOS uses clean instrument air (i.e., air with a dew point of negative 40°) to lift oil from the auxiliary reservoir shown in Fig. A.2.1. Oil mist created in a simple stationary-vane vortex generator (see Fig. 6.1) is then routed to standard spray nozzles or to reclassifiers that direct the spray (or oil mist) into the rolling element bearings. Air and entrained oil travel through the bearings and into the coalescer. Gravity flow causes the disengaged oil to either fall or flow back into the oil sump and reservoir, while the carrier air, still under slight positive pressure, leaves the coalescer relatively oil-free. The coalescer can be designed with a small lid through which makeup oil could be added. It is threaded into the housing top where pumps normally have a vented filler cap. The vent cap can be discarded, since it is no longer needed.

How ADIOS works can also be visualized from Fig. A.2.1, a schematic that depicts its functionality. It should be noted that a well-designed coalescer is an important component. It needs to be porous enough to allow the oil mist pressure in a bearing housing to dissipate after 10, 20, or 30 s, at which time the sensing line signals the air shutoff valve to open. This will allow another blast of air to aspirate lube oil from the 2-gal reservoir. However, if the coalescer is too porous, the air expelled through it into the atmosphere would contain too many of the small, atomized oil globules that are carried in a blast of air, which we call oil mist. Some of the mist will have been converted into liquid oil by the turbulent action in the rotating bearings; the remainder will migrate into the coalescer. An optimized coalescer design will likely return about 99% of the oil entrained in the mist leaving the bearings. Oil is collected and reused in the closed ADIOS system.

A.2.4 Components making up ADIOS

In the pump cross section of Fig. A.2.1, clean, dry instrument air enters upstream of a small pressure regulator. The reduced pressure air leaving the regulator goes into a suitably configured on–off valve. A sensing line connects this valve with the interior space of the pump's bearing housing. Actuation of the on–off air supply valve is triggered by the sensing line, realizing that the normal pressure inside the bearing housing (say, 35″ of water column) has decayed to 10″ or 7″ of water column. This decaying will occur because slightly pressurized air will flow into the coalescer and escape into the atmosphere. Tiny oil droplets will begin coating the inside of the coalescer and soon combine into large drops of oil. These large drops will fall back and return to the 2-gal (7.5 L) reservoir or bearing housing sump.

As air escapes into the atmosphere while the on–off valve remains closed, the sensing line tells the pilot valve that we are now at 7″ or 10″ of water column. Somewhere between 7 and 10, the on–off valve opens, and the cycle repeats. Cycle times are a function of coalescer size and porosity. A small cone-tipped setscrew allows trimming the air flow near the junction of the sensing line and on–off valve; cycle times are adjustable and probably cover the mid-range of a component combination, from perhaps 4 to 60 bursts per minute.

Suffice it to say that a number of component styles and types could be used to fully configure ADIOS. This is especially important when placing miniaturized ADIOS units inside a bearing housing or when flanging such a unit onto a bearing housing is being considered by machinery manufacturers.

Once the on–off valve opens, air is admitted to the small venturi or vortex generator located as shown in Fig. A.2.1 or inside the bearing housing. The venturi or vortex generator pulls oil into the air stream; the air–oil mixture may have the consistency and appearance of oil mist, also known as oil fog. Either liquid oil or oil fog is acceptable for flowing through small nozzles into rolling element bearings. In general, these nozzles would be mounted between a bearing housing protector seal (see Fig. 6.21) and the adjacent sides of the thrust and radial bearings. Two bearing housing protector seals are needed per bearing housing. Whether oil mist or a short blast of liquid oil (i.e., an oil jet) arrives at the bearing is determined by the geometries of the vortex generator or venturi. Either of the two would make up the oil mist generator, or "OMG." The downstream spray or reclassifier nozzles are either separate components or are passageways of suitable diameter (see Tab. 6.1) drilled into the end walls or end caps of a bearing housing.

A.2.5 Other options, explained

Users would find the ADIOS approach especially attractive for remote-mounted pumps and stirrers, perhaps in a tank farm or at other off-site locations. However, ADIOS units are by no means restricted to process pumps; agricultural and mining machinery come to mind immediately. Rail and other transportation equipment will have ample opportunities to consider ADIOS.

If the commercial developer's experiments show that the desired two drops of oil are entrained in a blast and deposited at a rate of one blast per minute, it would take five days for one liter of oil to go through each of the two bearings, namely, through the more highly loaded thrust bearing and the generally lightly loaded radial bearing. (The reader may recall that 7,200 drops of oil equal one liter of oil.) We project that approximately 99% of the mist which coalesced after traveling through bearings will have coalesced or will reach the coalescer. Little, if anything, will escape through a highly effective coalescer into the environment.

One calculation assumed that 10 mL of oil would be lost in the span of 5 days, equaling 710 mL/year. One could enlarge the sump volume by threading a cylindrical or rectangular 1-L stainless steel extender volume into the bottom drain port of a bearing housing. As in thousands of other instances, the sky is the limit for innovative developers.

As a point of reference, the two oil mist handbooks [34, 35] give 6 mL as an estimate of oil entrained in air and leaving as spent mist per hour from an average-sized process pump. Using closed oil mist systems reduces that old projection by orders of magnitude. But the following establishes the correctness of a 50-year-old guideline: Only a miniscule amount of oil will actually be consumed in an average 3-inch diameter bearing: Per Tab. 6.1, this 3 BI bearing requires 5.4 SCFH of oil mist. Page 28 in [34] mentions that 0.054 cubic inches of oil = 0.9 ml/h are carried through the bearing.

As we recall that 99% of the oil will be returned as coalesced liquid, we realize the appropriateness of the term "miniscule."

Bearings will not overheat so long as an oil film separates the rolling elements from their inner and outer raceways. Much heat is created when bearing elements must overcome the frictional resistance of lots of liquid oil in a conventionally lubricated bearing. Oil mist lubricated bearings typically run 18 °F (10 °C) cooler than equivalent bearings plowing through lots of oil. There will be little friction with the ADIOS approach and while opting for an ISO VG 68 PAO/diester or lesser viscosity lube oil mixture.

Appendix III: Jobsite receiving and protection

A.3.1 API RP-686

The Recommended Practice for Machinery Installation and Installation Design – API RP-686 has gained in importance over the last 20 plus years, providing checklists for the installation and pre-commissioning of new and existing machinery. Since its first edition in April 1996, and with the improved second edition in December 2009, this specification has now been adopted and accepted globally for use as an integral part of job specs for many major grassroots projects, end-users, EPC (engineering procurement contractors), and OEMs (original equipment manufacturers).

The intent of these API-RP-686 checklists is to document, recommend, and facilitate proper installation and commissioning, and to prevent or minimize startup delays. This ultimately provides end-users with reduced life cycle cost.

In itself, the API RP-686 specification provides enough details and scope to be used as a contractual document between end-user and EPC. In particular, the section pertaining to "**Jobsite Receiving and Protection** – **Chapter 3,**" is increasingly used by the EPC and OEM in their contracts. This mutually agreed contractual document acts as a guide on how the equipment should be handled during receipt and deals with general handling of the equipment during storage.

The "**Jobsite Receiving and Protection**" document is usually written by the OEM, reviewed, and signed off by the EPC/owner-purchaser. Many elements of what the reader will have found in the main chapters of our "Optimized ..." book are woven into modern OEM/EPC/owner-purchaser documents as a written narrative.

The written narrative tries to meet the intent of Chapter 3 of API RP-686 and, in most cases, supersedes it by ascertaining that any gaps in this chapter of the specification are recognized and closed. It is the opinion of the author that many potential litigation risks between EPC and OEM would have been offset by having (and following) a clearly written "**Jobsite Receiving and Protection**" document.

However, protection of the equipment commences prior to leaving the OEM facility, and although API RP-686 does not address this phase of a project, it is important for the EPC and the OEM to have clarity on the following questions:

- How will the equipment be packaged for shipment? Examples: wooden crates with internal supports or canister/containers with N_2 blanket?
- Will the equipment arrive at site fully preserved? If so, how?
- Were onsite conditions considered – for example humidity, heat, and cold– for the type of preservative used?
- Will preservation be for short-term or long-term storage?
- If extended storage time is required later due to slippage of project schedule, what are the backup plans for safe and continued storage and preservation?

https://doi.org/10.1515/9783110749441-020

An inspector, normally employed by the EPC to oversee the job, should be acquainted with the specification, and therefore should know the answers to the above questions prior to final packaging. He represents the EPC and should be the one signing off on packaging, on their behalf, just before the equipment leaves the OEM's facilities.

All equipment and materials shipped loose should be packed and preserved in a manner suitable for land and sea transport. The preservation method selected must resist damage due to environmental effects – such as corrosion – and should last the entire period given in the warranty clause of the contract.

All packages should be prepared for transit in accordance with the applicable OEM recommended specifications. Each equipment item –such as compressors, steam turbines, gearboxes, motors, and pumps – should have distinct packing specifications specific to the type of equipment. This must be verified by the inspector representing the EPC. However, a final cursory inspection is sometimes done by the inspector, following the guidelines given in Tab. A.3.1.

A.3.2 Jobsite receiving and inspection

Prior to receiving the shipped equipment and material, the EPC's construction manager must ascertain that procurement schedules, OEM storage documentation, installation manuals and drawings are forwarded to the OEM's and owner-purchaser's designated machinery representatives. The OEM's team receiving the shipment should review weights and method of shipping to determine the type of equipment required to unload the shipment, and make plans accordingly.

It is highly recommended that a receiving team comprising the OEM representative, the EPC-designated representatives, and the end-user engineers (machinery, instrument, and electrical) is made available to receive the shipment.

Using Annex B of API RP-686 – Machinery Receiving and Protection Checklist, each individual member of the team makes his/her own assessment, following the guidelines given in Section 2.2 ("a" through "o") of RP-686, Chapter 3. The team members will then consolidate their findings and make recommendations regarding the state and conformity of the shipment.

Findings of any irregularities in the shipment such as external damage of the packing or impact marks must be reported and followed up by a thorough inspection inside. However, if aluminum protective foils are used (protection from temperature and humidity), the foil must not be cut open until unpacking. Any damage found during the inspection must be reported immediately to the OEM, the applicable transport organization, and the insurance company. Late claims are generally not accepted.

A.3.3 Jobsite storage and protection

Storage and protection care recommendations by the equipment manufacturer must be strictly followed by the EPC and end-user representatives for site storage and protection. It is noteworthy that failure to follow the equipment manufacturer's guidelines for storage and protection care may void the warranty. Realize, however, that "Manufacturer's Guidelines" may well be a document that was compiled, accepted, and issued by the three parties OEM, EPC, and owner-purchaser.

A thorough review of the procurement document must be done to determine if the equipment had been prepared and preserved for a predetermined period. Failure to do so could result in improper storage and preservation procedures at site. As an example, if API RP-686 was used as the governing document for jobsite protection and preservation, then the equipment preparation for shipment would be good only for six months of outdoor storage from the time of shipment, with no disassembly required before operation, except for inspection of bearings and seals.

Whenever the manufacturer's recommendations are not available, then – and only then– can API RP-686 be used as a minimum acceptable guide. Jobsite preservation and storage guidelines for specific equipment such as pumps, blowers, fans, compressors, instrumentation, and steam turbines are addressed in Sections 2.3–2.20 of API RP-686. Annexes B, C and D are used to log compliance of the items in each of the checklists.

The EPC construction manager usually designates a team to execute the equipment storage and protection program, once equipment gets to the laydown yard. The team is usually led by an engineer, and it is his or her responsibility to give a status report on equipment protection to client engineers (essentially the owner-purchaser) at a weekly meeting. Client engineers attending the weekly meeting represent several disciplines: fixed and rotating mechanical engineers; electrical; instrument and controls engineers (all elsewhere called "SMEs").

Table A.3.2 represents a typical jobsite protection program issued by a competent rotating equipment manufacturer. The protection program starts from the time equipment arrives at the laydown yard, prior to installation, and extends through equipment installation, and beyond. It ends with handover of an entire project.

Thus, equipment protection continues after equipment installation, as can be seen in Tab. A.3.2. However, the EPC generally consolidates the equipment protection program from each of the equipment manufacturer into a spreadsheet format for each area of the plant. Tables A.3.3 and A.3.4 show an example of what needs to be done, area-by-area, for the protection of each type of rotating equipment, once installed.

The weekly preservation meeting continues even after equipment installation; however, a walkthrough in each area of the plant is usually done following the meeting.

Tab. A.3.1: Checklist used by inspector prior to shipping the equipment.

Item #	Items to be checked	Requirements	Inspector's comments
1	Preservation and protection	Items subjected to long-term storage must conform to the applicable specification recommended by the OEM or Tab. 1, Ref. [1] or Annex A, API RP-686	
2	Exposed machined surfaces	External machined surfaces exposed to corrosion must be coated with an appropriate "steelguard" coating such as CORTEC VCI 389 or equivalent	
3	Flanged openings	Openings must be closed with full-face covers made from soft timber or plastic (hardboard or similar materials are not permitted). Before covering the openings, the surfaces must be treated similar to item#1.	
4	Threaded connections	Male and female threaded connections must be treated with a de-watering type protection fluid, then plugged with metal cap or plug.	
5	Instrument and control panels	Instruments liable to be damage must be removed and shipped separately. These instruments must be individually packed and labeled. Required packing can be enclosed with polyethylene or aluminum foil with favorable amounts of desiccants securely fastened within the package. All panels shall be cushioned to prevent damage. Gauges and switches that have glass or acrylic facings must be protected. Plywood blanks can be used but must be securely taped in position.	
6	Skid assemblies	Partial protection of skid assemblies is required by bolting timber packings to the underside of the skid.	
7	Crating/casing	The construction of the crate or casing must be suitable for mechanical handling and meet the specification previously agreed to by OEM and EPC/purchaser.	
8	Identification and markings	Consignments must be marked in accordance with the client's specification and include dimensions and total weight of the unit. Warning labels must be attached to ensure receiver is aware of protection measures and actions taken prior to storage and startup.	

Tab. A.3.2: Manufacturer's recommended preservation and protection procedure at site.

Equipments (Type & Procedure of Storage / Time of Storage)		Before Installation: Within 1 Year		Before Installation: Over 1 Year		Under Installation & After Completion of Installation: Common items for After Installation To Commissioning Operation									Over 1 Month to Within 1 Year		After Trial Operation		
		Indoor Storage	Outdoor Storage	Partial Overhauling	Adding Repairing of Rust Preventive and Drying Agent	Checking of Awning Sheets (Every day)	Rotation of Rotor (Once a Week)	Blinding of Opening Areas	N_2 or Inert Gas Sealing	Adding Repairing of Rust Preventive and Drying Agent	Checking of Drain (Once a Week)	Circulation of Oil (Every day)	Partial Overhauling (Once a Year)	Using Space Heater To Prevent Moisture	Making By-pass Line of Seal Oil Line	N_2 or Inert Gas Sealing	Drying	Adding of Volatile Rust Preventive Agent	Cleaning after Overhauling
Main Machine	Turbine	Δ	O	O	O	O	Δ	O	Δ	O	O	O	O			O	O	O	O
	Compressor	Δ	O	O	O	O	Δ	O	Δ	O	O	O	O		O⑤	O⑤	O	O	O
	Main Motor	Δ	O	O	O	Δ	Δ			O			⑥	⑦					O
	Gear	Δ	O	O	O	O	Δ	O	Δ	O	O	O	O			O	O	O	O
Apparatus Auxiliary Equipment	Oil Console	Δ	O	O	O	Δ		O	Δ		Δ	O	O						O
	Oil Reservoir	Δ	O	O	O	Δ		O	Δ	O	Δ	O	O			Δ			O
	Oil Pump(s)	Δ	O	O	O	O	Δ	O	Δ	O	O	O	O		⑤	O⑤			O
	Electric Motor(s) – Outdoor Type	Δ	O	O	O	Δ	Δ			O			⑥						O
	Electric Motor(s) – Indoor Type	O		O	O	O	Δ			O	③		⑥				②		O
	Electric Oil heater	Δ	O	O	O			O		O			O						O
	Gland Condenser (s)		O	O	O			O	Δ	O	Δ		O			Δ			O
	Head Tank (s)		O	O	O			O	Δ	O	Δ		O			Δ			O
Instrument	Governor (s)	O		O	O	O		O		O	②	O	O		O		②		
	Pneumatic	O		O	O	O		O	④	O			O				②		
	Electrical	O		O	O	O		O		O	②		O				②		
	Others	Δ	O	O	O	O		O		O			O				②		
Piping	Oil		O	O	O			O	Δ	O		O	O		O				O
	Steam		O	O	O			O	Δ	O			O			O			O
	Gas		O	O	O			O	Δ	O	O		O			O			O
Spare Parts	Instrument(s)	O		O	O								O						
	Others	Δ	O	O	O								O						
Others	Insulation(s)	O		O							③		O						
	Welding Rod(s)	O		O							③		O						
	Sealant Compound	O①			O①								O						

Explanation of Mark

O To be performed absolutely
Δ To be performed as much as possible

Specific Notices

① Store in a dark & cool place and be sure to make note of its storage period or shelf life.

② Remove and store indoors except for during installation and/or operation.

③ Store indoors and is not to be exposed to the rain.

④ Always feed dry instrument air or nitrogen gas after completion of installation.

⑤ Do not remove rust preventive grease in the seal portion before commencement of oil circulation.

⑥ Perform an insulation resistance test every 3 months.

⑦ Space heater shall be used under not running.

		Mechanical Equipment			
				AREA 17	
Nos.	Equipment Tag Nos.	Discription	QTY	Frequency	Preservation Requirements
1	UE SGF AN01	FD FAN	1	24 weeks	1. Inspect service coatings and repair as necessary. Use Cortec VpCI368 for
2	UG SGF AN01	FD FAN	1		machined bare surfaces.
				weekly	2. Energize motor condensation heaters. MV motors only . (> = 160Kw)
				daily	3. Monitor continous power supply to condensation heaters.(MV motors
				24 weeks	4. Spray motor bearings with protective oil layer, Tectyl 511 or equivalent
					through oil hole. MV motors only (> = 160Kw)
				24 weeks	5. Check motor winding resistance of MV motors.
				10 weeks	6. Rotate Motor Shaft 10–1/4 turns.
				4 weeks	7. Rotate Fan Shaft 10–1/4 turns.
				weekly	8. Rotate Turbine Shaft 2–1/4 turns.
3	ULP01-A	LP/HP Boiler Feed Water Pump	1	24 weeks	1. Inspect Flange Covers, Caps etc.
4	ULP01-B	LP/HP Boiler Feed Water Pump	1	24 weeks	2. Inspect service coatings and repair as necessary. Use Cortec VpCI368 for
5	ULP01-C	LP/HP Boiler Feed Water Pump	1		machined bare surfaces.
6	ULP01-D	LP/HP Boiler Feed Water Pump	1	weekly	3. Rotate Turbine Shaft 2–1/4 turns.
7	ULP01-E	LP/HP Boiler Feed Water Pump	1	24 weeks	1. Inspect Flange Covers, Caps etc.
8	ULP01-F	LP/HP Boiler Feed Water Pump	1	24 weeks	2. Inspect service coatings and repair as necessary. Use Cortec VpCI368 for
9	ULP01-G	LP/HP Boiler Feed Water Pump	1		machined bare surfaces.
				weekly	3. Energize and inspect motor condensation heaters. MV motors only.
				daily	4. Monitor continous power supply to condensation heaters. MV motors only .
				24 weeks	5. Spray motor bearings with protective oil layer, Tectyl 511 or equivalent
					through oil hole.
				10 weeks	6. Rotate Motor Shaft 10–1/4 turns.
10	ULP04-A	Suspect Condensate Pump	1	24 weeks	1. Inspect Flange Covers, Caps etc.
11	ULP04-B	Suspect Condensate Pump	1	24 weeks	2. Inspect service coatings and repair as necessary. Use Cortec VpCI368 for
					machined bare surfaces.
				weekly	3. Energize and inspect motor condensation heaters. MV motors only.
				daily	4. Monitor continous power supply to condensation heaters. MV motors only .
				24 weeks	5. Spray motor bearings with protective oil layer, Tectyl 511 or equivalent
					through oil hole.
				10 weeks	6. Rotate Pump Shaft 10–1/4 turns.
				10 weeks	7. Rotate Motor Shaft 10–1/4 turns.

Tab. A.3.3: Activities related to installed machinery preservation and storage.

Tab. A.3.4: Installed machinery preservation and storage (contd.).

				AREA 18	
Nos.	**Equipment Tag Nos.**	**Discription**	**QTY**	**Frequency**	**Preservation Requirements**
1	UWP-01-A	Cooling Water Supply Pump	1	24 weeks	1. Inspect Flange Covers, Caps etc.
2	UWP-01-B	Cooling Water Supply Pump	1	24 weeks	2. Inspect service coatings and repair as necessary. Use Cortec VpCI368 for
3	UWP-01-C	Cooling Water Supply Pump	1		machined bare surfaces.
4	UWP-01-D	Cooling Water Supply Pump	1	12 weeks	3. Connect pump to Oil Mist. (Check & refill oil mixture of 5 parts VG46 to 1
5	UWP-01-E	Cooling Water Supply Pump	1		part of Cortec 329)
				12 weeks	4. Connect turbine to Oil Mist. (Check & refill oil mixture of 5 parts VG46 to 1
					part of Cortec 329)
				12 weeks	5. Connect gearbox to Oil Mist. (Spray with oil only, mixture of 5 parts VG46
					to 1 part of Cortec 329)
				10 weeks	6. Rotate Pump Shaft 10–1/4 turns.
				10 weeks	7. Rotate Lube Motor Shaft 10–1/4 turns.
				weekly	8. Rotate Turbine Shaft 2–1/4 turns.
6	UWPM-01-D	Cooling Water Supply Pump Motor	1	24 weeks	1. Spray motor bearings with protective oil layer, Tectyl 511 or equivalent
7	UWPM-01-E	Cooling Water Supply Pump Motor	1		through oil hole.
8	UWP ST-01-A	Steam Turbine	1	4 weeks	1. Introduce small quantity of VG46 oil through bearing breather or filler plug and
9	UWP ST-01-B	Steam Turbine	1		turn rotor 2-3 times to distribute oil around the journal and bearing surface.
10	UWP ST-01-C	Steam Turbine	1		
11	UWX P01-A	Lube Oil Console	1	12 weeks	1. Cleaning and covering of console panel.
12	UWX P01-B	Lube Oil Console	1		
13	UWX P01-C	Lube Oil Console	1		

References

[1] Eschmann, P; Hasbargen, L. and Weigand, K. *Ball and Roller Bearings: Theory, Design and Application*. Hoboken: John Wiley & Sons, 1985.

[2] Bloch, H. P. "Ranking Lube Applications in Pumps and Electric Motors." *Proceedings of the 30th Texas A&M University International Pump Users Symposium*, 2014.

[3] SKF USA. "MRC Engineering Handbook", Kulpsville, PA, "MRC Engineering Handbook", Publication M190-730, 1993.

[4] SKF USA. "MRC Bearing Solutions for the Hydrocarbon Processing Industry", Kulpsville, PA; Revised Publication M230-710, 1996.

[5] SKF USA. "MRC Bearings for Pumps", Kulpsville, PA; "MRC Bearings for Pumps", Original Publication M230-710, 3/91 ABG, 1991.

[6] MRC Bearing Services; Jamestown, NY 14701; "MRC Extra-Wide Ball Bearing Retrofit Kit," Publication M212-401 1991.

[7] Bloch, H. P. and Perez, R. X. *Pump Wisdom:* "Essential Centrifugal Pump Knowledge for Operators and Specialists." Revised and expanded, 2nd edition. Hoboken, NJ: Wiiley & Sons, 2021, ISBN 978-1-118-04123–9.

[8] Bloch, H. P. *Petrochemical Machinery Insights*. Oxford, UK and Cambridge, MA: Elsevier Publishing Company, 2017, (ISBN 978-0-12-809272–9).

[9] SKF Bearing Maintenance and Replacement Guide, Catalog #3600, 1986.

[10] Bloch, H. P. and Budris, A. *Pump User's Handbook: Life Extension,* Fourth Edition. Lilburn: Fairmont Publishing Company, 2013.

[11] Bloch, H. P. and Lee, B., "Breaking the Cycle of Pump Repairs," Proceedings of the 30th Texas A&M International Pump Users Symposium, 2014.

[12] Baudry, R. A. and Tichvinsky, L. M. "Performance of Oil Rings." *Journal of Basic Engineering*, Volume 82, 1960, 327–334.

[13] Bloch, H. P. Confirming consulting report for a Texas-based smelter, July 30, 2009. Excerpts published in the article, "Deferred Maintenance Causes Upsurge in BFW Pump Failures." Hydrocarbon Processing, May 2011.

[14] Bloch, H. P. *Fluid Machinery: Life Extension of Pumps, Gas Compressors and Drivers*. Berlin/ Germany: DeGruyter, 2020, ISBN 978-3-11-067415–6.

[15] Bloch, H. P. and Bannister, K. *Practical Lubrication for Industrial Facilities, Third Edition*. Lilburn: Fairmont Press, *2016*.

[16] Bradshaw, S., et al., "Factors Affecting Oil Ring and Slinger Lubricant Delivery and Stability," 2014, Proceedings of the Texas A&M International Pump Symposium, Houston, Texas.

[17] Wilcock, D. F. and Booser, E. R. *Bearing Design and Application*. New York, NY: McGraw-Hill Company, 1957.

[18] Heshmat, H. and Pinkus, O. "Experimental Study of Stable High-Speed Oil Rings." *Journal of Tribology*, Volume 107, No. 1, January 1985, 14–22.

[19] Bloch, H. P. *Improving Machinery Reliability*, 3rd Edition. Houston, Texas: Gulf Publishing Company, 1982/1988/1998, (ISBN 0-88415-661–3).

[20] SKF Publication 100–955, 2nd Edition, "Bearings in Centrifugal Pumps," Application Handbook, 1995.

[21] Bloch, H. P. "Oil mist lubrication: Is it justified and how should it be executed in the 1990s?" Hydrocarbon Processing, October 1990, pp.25

[22] Bradshaw, S. "Investigations into the Contamination of Lubricating Oils in Rolling Element Bearing Assemblies," Proceedings of the Texas A&M International Pump Symposium, Houston, Texas, 2000.

https://doi.org/10.1515/9783110749441-021

[23] Bloch, H. P. "Power End Upgrades Will Avoid Pump Failures." *Proceedings of the 27th Texas A&M University International Pump Users Symposia*, 2011.
[24] Cantley,R. E. "The Effects of Water in Lubricating Oil on Bearing Fatigue Life." *ASLE Transactions*, Volume 20, No. 3, 1977.
[25] Schatzberg, P. "Influence of Water and Oxygen in Lubricants on Sliding Wear." *Lubrication Engineering*, Volume 26, 9, 1970.
[26] Schatzberg, P. and Felsen, I. M. "Effects of Water and Oxygen During Rolling-Contact Lubrication." *Wear*, Volume 12, 1968.
[27] Bloch, H. P. "Large Scale Application of Pure Oil Mist Lubrication in Petrochemical Plants." The American Society of Mechanical Engineers: ASME Paper No. 80-2/Lub-25, 1980.
[28] Bloch, H. P. and Geitner, F. K. *Practical Machinery Management for Process Plants – Maintenance and Repair, Volume 3, Third Edition*. Oxford: Gulf Professional Publishing, 2005.
[29] Bloch, H. P. and Lee, B. "Breaking the Cycle of Pump Repairs," Proceedings of the 30th Texas A&M International Pump Users Symposium, 2014.
[30] Ehlert, D. "Consider Closed-Loop Oil Mist Lubrication." Hydrocarbon Processing, June 2011: https://www.hydrocarbonprocessing.com/magazine/2011/june-2011/special-report-processplant-optimization/consider-closed-loop-oil-mist-lubrication.
[31] Clapp, A. M. and Wilcox, F. B. "Plant Lubrication." *Proceedings of the Seventh Texas A&M University International Turbomachinery Symposium*, 1978.
[32] Miannay, C. R. "Improve Bearing Life." Hydrocarbon Processing, May 1974.
[33] Bloch, H. P. "Dry Sump Oil Mist Lubrication for Electric Motors." Hydrocarbon Processing, March 1977.
[34] Bloch, H. P. *Oil Mist Lubrication Handbook,* First Edition. Houston: Gulf Professional Publishing Company, 1987.
[35] Bloch, H. P. and Shamim, A. *Oil Mist Lubrication: Practical Applications*. Lilburn: Fairmont Press, 1998.
[36] Ehlert, D. "Lubrication Maintenance Planning." Pumps & Systems Magazine, 2013.
[37] Bloch, H. P. "Oil Mist Lubrication Cuts Bearing Maintenance." Plant Services Magazine, November 1983.
[38] Ehlert, D. and Williams, M. "Oil Mist Lubrication." Pump Industry Magazine, 2013.
[39] MRC. *MRC Engineer's Handbook*. Bearing General Catalog 60, Second Edition, 1982, 197.
[40] Bloch, H. P. and Ehlert, D. "Reliability Milestones Reached by Plant-Wide Oil Mist Systems." *Uptime Magazine*, 2013: https://reliabilityweb.com/articles/entry/reliability_milestones_reached_by_plant-wide_oil_mist_systems.
[41] Bloch, H. P. "Best of Class Lubrication for Pumps and Drivers." Pumps & Systems, April 1997.
[42] Bloch, H. P. "Mixing of Greases – Compatibility," Hydrocarbon Processing, March 2017.
[43] Bloch, H. P. "Applying Oil Mist." Fluid Power Journal, February 2005.
[44] ExxonMobil Lube Marketing Bulletin/Newsletter on Grease Guidelines.
[45] SKF General Catalogue 5000E, June 2003.
[46] Ehlert, D. "Getting the Facts on Oil Mist Lubrication." *Proceedings of the Texas A&M Middle East Turbomachinery Symposium*, 2011.
[47] Ehlert, D. "Oil Mist Lubrication in the Hydrocarbon Processing Industry." Machinery Lubrication, 2011.
[48] Bloch, H. P. and Geitner, F. K. *Practical Machinery Management for Process Plants – Major Process Equipment Management and Repair, Volume 4, Second Edition*. Oxford: Gulf Professional Publishing, 1997.
[49] Ehlert, D. "Eliminating Risks in Bearing Lubrication." World Pumps Magazine, 2012.
[50] Bloch, H. P. and Ehlert, D. "Get the Facts on Oil Mist Lubrication Justification." Hydrocarbon Processing Magazine, 2008.

[51] Ehlert, D. "Oil Mist: The Ultimate Oil Filter." Pump Engineer Magazine, 2013.
[52] Bloch, H. P. "Preservation by Oil Mist Application." Plant Services Magazine, November 1987.
[53] Bloch, H. P. "Oil Mist Lubrication for Electric Motors." Hydrocarbon Processing, August 2005.
[54] Bloch, H. P. and Geitner, F. K. *Machinery Uptime Improvement*. Oxford: Butterworth-Heinemann, 2006.
[55] Ehlert, D. "Centralized Lube Systems." Plant Services Magazine, 1991.
[56] Ehlert, D. "Bearing Lubrication Trends and Tips." Pumps & Systems Magazine, 1993.
[57] Ehlert, D. "Lubrication Made Simple." Pumps & Systems Magazine, 2005.
[58] Ehlert, D. "Consider Closed-Loop Oil Mist Lubrication." Hydrocarbon Processing Magazine, 2011.
[59] Bloch, H. P. and Ehlert, D. "Get the Facts on Justifying Oil Mist Lubrication," Hydrocarbon Processing Magazine, 2008.
[60] Ehlert, D. "Consider Updating Your Lubricant System." Hydrocarbon Processing Magazine, 2012.

Index

https://doi.org/10.1515/9783110749441-022

About the author

Illustration: At the projection screen (Des Moines, Iowa, 2010), the author is teaching visiting overseas senior reliability managers when, where, and how to implement upgrade strategies in their respective home countries.

Heinz P. Bloch resides in Montgomery, Texas. Before commencing his engineering career in 1962, he absolved an apprenticeship as a low-voltage technician and worked as an experimental machinist. He was employed as a design engineer by Johnson & Johnson, followed by Esso/Exxon Research, and retired as Exxon Chemical's regional machinery specialist for the United States. He has authored or co-written over 790 publications, among them were 24 comprehensive books on practical machinery management, failure analysis, failure avoidance, compressors, steam turbines, pumps, and optimized lubrication for industry. He holds BS and MS degrees (cum laude) in mechanical engineering from the New Jersey Institute of Technology's Newark College of Engineering ("NCE"). In 2019, he was selected to be one of 10 inaugural inductees into NCE's "*Top 100 Hall of Fame*," which honors its most distinguished alumni.

https://doi.org/10.1515/9783110749441-023

www.ingramcontent.com/pod-product-compliance
Lightning Source LLC
Chambersburg PA
CBHW080933240926
51849CB00003BA/440
* 9 7 8 3 1 1 0 7 4 9 3 4 2 *